Studies in applied regional science

Vol. 14

Editor-in-Chief

P. Nijkamp
Free University, Amsterdam

Editorial Board

Å.E Andersson
University of Gothenburg, Gothenburg
W. Isard
Regional Science Institute, Philadelphia
L.H. Klaassen
Netherlands Economic Institute, Rotterdam
I. Masser
State University, Utrecht
N. Sakashita
Osaka University, Osaka

Studies in applied regional science

This series in applied regional, urban and
environmental analysis aims to provide
regional scientists with a set of adequate tools
for empirical regional analysis and for practical regional planning problems. The major
emphasis in this series will be upon the
applicability of theories and methods in the
field of regional science; these will be presented in a form which can be readily used
by practitioners. Both new applications of
existing knowledge and newly developed ideas
will be published in the series.

To
Trudi

Multiobjective regional energy planning

Application to the energy park concept

Peter D. Blair
University of Pennsylvania

Springer-Science+Business Media, B.V

ISBN 978-94-017-2371-8   ISBN 978-94-017-2369-5 (eBook)
DOI 10.1007/978-94-017-2369-5

Library of Congress Cataloging in Publication Data

Blair, Peter D
  Multiobjective regional energy planning.

  (Studies in applied regional science ; v. 14)
  Includes index.
  1. Energy Policy--Mathematical models. 2. Energy
facilities--Planning--Mathematical models. 3. System
analysis. I. Title. II. Title: Energy park.
III. Series.
  HD9502.A2B62   333.7   78-12282
  ISBN 978-94-017-2371-8

Preface

In recent years, the scope of energy planning has been broadened to include a variety of additional considerations such as socioeconomic and environmental impacts. The fundamental purpose of energy planning is to formulate policy. Policy must be formulated in response to the interests which that policy would affect. A planning model called policy programming is developed in this work from basic concepts of hierarchical system theory and input-output analysis. The model is used in planning for energy park development in a specific region.

I wish to acknowledge gratefully the suggestions of Thomas L. Saaty and Ronald Miller who commented at length on various drafts of the manuscript. Support for this work was provided in part by the U.S. Energy Research and Development Administration, the U.S. Federal Energy Administration, and the University of Pennsylvania Energy Center.

Peter Blair
December, 1977

Contents

Part One

Systems theory and energy planning

1. Introduction

1.1. ENERGY PLANNING

Traditional energy planning problems have usually considered a single high priority objective such as minimizing costs, maximizing availability, or ensuring reliability. In recent years, however, this traditional view has been re-evaluated so that lower priority objectives such as maintaining environmental quality or minimizing resource depletion have been elevated to equal or greater importance than that of the primary objective. Such objectives were formerly treated either as constraints or neglected entirely.

This new perspective on energy planning has prompted the development of a myriad of 'multiobjective' decision-making tools.[1] Unfortunately, many of these new tools are restrictive, particularly in long-range energy planning problems, the reasons for which are twofold:

1. A principal purpose of long-range energy planning tools is to formulate energy policy. The evolution of energy policy and its implementation are affected by a highly diverse collection of 'policy-making interests' (industry, consumers, energy suppliers, etc). The collective objectives of these policy makers are not effectively captured by current methodologies.
2. The activities in an energy system such as industrial production, pollution emission and energy consumption are interdependent. The links between the individual objectives of policy-making interests and these interdependent activities have not been adequately defined in current methodologies.

---

1. For surveys of multiobjective tools see Johnsen [204], Cohon and Marks [182], and Cochrane and Zeleny [18], or more recently Nijkamp and van Delft [237, 238], or Nijkamp and Rietveld [239].

The relationships among activities in an energy system constitute a set of *system process functions* that regulate flows in an energy system (energy, pollution, dollars, materials, etc). Of principal concern here will be the development of a method for relating the objectives of a number of policy-making interests to a model of these system process functions. The methodology will be concerned primarily with large-scale, long-range, energy planning problems, the consideration of which has just recently surfaced to become an important element in formulating and analyzing energy policy.[2] Examples of such problems are contingency planning for energy shortfalls, transcontinental pipeline development, major energy research and development program development, and the recently proposed energy park concept.

Complexity is the overriding problem one encounters in studying a system influenced by large-scale planning efforts such as those just described. These planning efforts should cut across a broad spectrum of issues: technological, environmental, social, political, legal, financial, etc. By complexity we mean that the factors affecting these issues form an intricate combination of elements that make the problem appear very difficult to examine as a whole. Current approaches to these problems have been limited in scope, dealing with one or two issues. As a result, there are today remarkably few operational methodologies that comprehensively address the wide range of issues involved in large-scale energy planning problems. A major reason for this appears to be the scarcity of methods that can be used to interrelate the components of such highly complex systems.

A key to dealing with the problem of complexity might be to arrange the system in a manner that reveals the interaction of its components. Perhaps the most powerful way of accomplishing this arrangement is to utilize a hierarchical structure. Whether one is concerned with understanding the actual structure and flows in a system or with the functional interaction of its components, a hierarchical model of the system enables one to study interactions between different levels of activities. The literature of hierarchies is large but there have been few attempts to operationalize hierarchical concepts.[3] Chapter 2 will examine some hierarchical systems concepts that can be utilized in structuring and analyzing energy systems that cut across the broad spectrum of issues mentioned above.

2. Palmedo [157] discusses recent large scale energy policy models.
3. Some general references are Pattee [47], Mesarovic et al. [44], Whyte et al. [58], Weiss [57]. The subject will be dealt with in more detail in chapter 2.

In energy planning one must be ultimately concerned with formulating and implementing policy. Policy should be developed in response to the parties or 'actors' that would be affected by that policy. Of principal concern then in energy planning is the interaction of these actors or policy-making interests. This kind of interaction can be expressed in a hierarchy where each actor has a number of objectives as well as an amount of influence that he can bring to bear in fulfilling those objectives. An important result of chapter 2 will be a framework, formulated from basic hierarchical concepts, that will serve to interrelate the individual objectives of policy-making interests as well as to compose their apparent collective objectives. This can be accomplished by first determining the relative preference that actors have for each of their objectives. Then the collective objectives can be composed by weighting the individual objectives according to the relative influence that each policy maker has in fulfilling his objectives.

A powerful method for operationalizing hierarchical concepts of influence and priorities on objectives is the recently developed eigenvalue prioritization method of Saaty.[4] This method will be briefly examined in chapter 2 and then modified in chapter 3 to relate the measures of influence and priority on objectives, that can be derived in the original model, to system variables in a model of system process functions in an energy planning system. The result will be a set of collective objectives or a *composite scenario* that describes a state to which actors in the system would collectively prefer that the system move.

Unfortunately, this composite scenario describes a state that *only* accounts for the interacting objectives and influence of policy makers. As a result, the scenario may not be feasible in terms of the system process functions that guide the flows in an energy planning system. For instance, the collective objectives of the actors may display preference for both a high level of industrial production and a low level of environmental pollution emission. There exists, however, a degree of interdependence between levels of industrial activity and levels of environmental pollution emission that is captured in the system process functions. Hence, the future collectively desired by the policy makers must obey the process functions of the system in order to project a feasible set of planning targets. The principal aim of chapter 3 will be to develop a method for composing a set of planning targets that is feasible in terms of a model of energy system process functions.

4. The general approach is given in Saaty and Khouja [53]. More comprehensive treatment is given in Saaty [49] and Saaty [51].

1.2. GENERAL APPROACH TO THE PROBLEM

A hierarchical model for determining the objectives of individual policy-making interests in a system of energy-related activities will be developed in chapter 3. This model will be modified from some basic hierarchical concepts of Saaty and will form the first stage of a multiobjective planning model that will be called *policy programming*. In application the result of this first stage of the policy programming procedure will be a *composite scenario* or a collection of factors, the values of which describe a future that is collectively preferred by the collection of policy-making interests being considered. This composite scenario is determined from the objectives and relative influence of individual policy-making interests.

The second stage of the policy programming procedure utilizes a model of system process functions for an energy system of industrial activities. This model is a set of input-output relationships among the industrial activities. A composite scenario that obeys the relationships given by the model of system process functions will be defined to be *consistent*. The purpose of the second stage of policy programming is to develop a *consistent composite scenario* given a composite scenario that was originally not consistent.

It will become clear in chapter 3 that there may be many such consistent composite scenarios. We seek one that is as close as possible to the original collectively preferred composite scenario. This will be accomplished by treating the original scenario as a set of planning targets. Each planning target can be traced back to a composition of objectives of individual policy-making interests coupled with the influence that each policy-making interest can bring to bear to fulfill its objectives. We recall that this composition was the manner in which the original planning targets were developed in the first stage.

The measures of influence and priority on objectives from the hierarchical model (first stage) will be used to assign priorities to these planning targets. The most desired objective of the most influential policy-making interest receives the highest priority.

These priorities give a sequence in which rigid adherence to planning targets might be relaxed in order to yield a consistent scenario. In other words, the path to consistency will be found by relaxing low priority planning targets (in reverse order of priority) until a consistent scenario results.

What is needed is a systematic method for composing the consistent composite scenario in the manner just described. Such a method is provided by goal programming, which is a mathematical programming procedure that seeks to minimize deviations from a set of established targets, subject to a constraint set. Invoking this procedure requires three items: (1) an analytical representation of the constraint set which, in the present case, consists of the technical relationships that control flows in the system – the system process functions; (2) a priority structure that gives the order in which planning targets may be relaxed; and (3) the collection of planning targets. We recall that the last two elements are exactly what result from the modified hierarchical model of the first stage. What we need then is a suitable analytical model of the system process functions.

Most current energy policy models have been restricted to developing detailed models for system process functions. One of the best developed of these models has been the generalization of Leontief's input-output model.[5] In chapter 3, a model of system process functions for use in policy programming will be developed from the basic tools of input-output. The goal programming problem can then be constructed using this model as the constraint set, and it will be solved by means of a modified simplex algorithm which seeks, subject to the constraint set, to achieve the highest priority planning targets first. When these targets have been achieved, the algorithm then attempts to fulfill lower priority targets without compromising the already established high priority targets. The result is a collection of factors (a scenario) in which some lower priority planning targets of the original composite scenario may not be achieved but is consistent with the model of system process functions.

1.3. PRINCIPAL SIGNIFICANCE

Policy programming provides a link between a model of system process functions, which is the form of most current energy policy analysis models, and the objectives of policy making interests in an energy planning system. The purpose of policy programming is to contribute to long-range energy planning and policy making in a manner that is sensitive to the multiple objectives of a number of policy-making interests. However, the principal problem to be reckoned with in this analysis is that results must conform with the technical characteristics or system process functions of an energy system.

Part Two of this study deals with the specific problem of planning for

5. See Leontief [149]. The generalization of input-output analysis is discussed in chapter 3.

energy park development. [6] The energy park problem has all the characteristics of large-scale planning problems that policy programming is meant
to address. Chapter 4 will deal with many of the specific considerations
peculiar to the energy park concept in some detail. Chapter 5 will characterize a target region where an energy park might eventually be developed and review a number of alternative planning options for development
in that target region.

We realize that we cannot hope to consider all of the aspects of the
energy park planning problem. Such an endeavor would be well beyond
the scope of this study. However, a number of issues currently of concern
in assessing overall feasibility of the energy park concept have to do with
the regional impacts of proposed energy park development. In particular,
the potential regional economic, environmental, energy, and employment
impacts of such a development are closely related to each other and are of
principal concern. Hence, whereas we cannot address many of the considerations in energy park development, we will be able to contribute to a
better understanding of these regional impact issues via application of the
policy programming framework. Limiting the problem in this manner
allows a hopefully more comprehensive treatment of some of the issues of
most concern in considering the energy park as a future alternative for the
electric power industry.

Chapter 6 will apply the policy programming procedure developed in
Part One to the specific energy park planning problem that will evolve out
of chapters 4 and 5. While the complete policy programming framework
could not be implemented due to computational restrictions, an abbre-
viated version was applied.

The relative success of policy programming in the specific case of analyzing energy park development options will be discussed in chapter 6. In
addition, use of the results in planning for implementation of an energy
park at the target site will be discussed. These results will be generalized
where possible to address the more general problem of overall feasibility of
energy parks at arbitrary sites. Finally, some limitations of the policy programming framework will be discussed along with their effect on validity
on results and some suggested improvements for overcoming these defi-
ciencies.

A very important result from the specific analysis of energy park development in this work suggests that regional agencies should take a much

6. An 'energy park' or 'energy center' is a large concentration of electrical generating capacity (10,000–50,000 megawatts) located at a single geographic site.

stronger role in planning large-scale energy facilities such as energy parks. Presently the complete responsibility of such planning efforts has rested with the electric utilities who are not well suited to deal with the host of regional impact issues that might accompany energy park development. Moreover, the results suggest that ultimate success of the energy park concept (at least in the study region) would require a carefully coordinated effort on the part of utilities, government and regional planning agencies.

The use of policy programming in dealing with the energy park problem is meant as a first step toward the more general problem of linking the objectives of policy makers that might be affected by a large-scale energy project to the factors that are controlled by these policy makers. Many problems have yet to be solved in the area of planning in large-scale energy systems; policy programming is meant to provide some insight into the nature of these problems.

2. Energy systems and planning

2.1. THE ENERGY PLANNING PROBLEM

2.1.1. The general theory of planning

The process of planning has been defined (Ozbekhan [13]) as a 'rationally organized, future-directed action system.' In recent years, the general theory of planning has undergone fundamental changes in the problems that it addresses. Many traditional planning problems such as optimal resource allocation, transportation network planning, or corporate planning, are stated with clear boundaries and can be shown to have explicit solutions. Many modern planning problems, on the other hand, such as regional planning or multi-model transportation planning are much broader in scope.

The boundaries of these modern planning problems often cannot be well defined since they involve such a wide range of factors, social, environmental, technological, political, and so on. In addition, the resultant plan must address the wide diversity of parties whose interests that plan affects.

An example of such a problem is the recently proposed energy park concept which will be dealt with in some detail in Part Two of this study. An energy park is a large concentration of electric generating capacity (10–48 GW_e) located at a single geographic site. Constructing and operating an energy park could have profound socioeconomic, environmental, political, financial, as well as many other impacts on the region in which it is sited.

Traditional planning methodologies that address well-defined systems and seek to achieve single, clear-cut objectives are not well suited to handling these modern planning problems. Hence, this unsuitability has prompted development of new methods for dealing with such problems. Ozbekhan observes: '... in the emerging methodology of planning one notices a definite change of focus, amounting to perhaps a fundamental shift in our world view, toward a systemic vision of reality.'

Ozbekhan's 'systemic vision' is, of course, a system approach to planning. This approach is seen in the relatively recent trend of planners attempting to orient planning methodology toward the complete system that their resulting plan would affect. This trend demands an answer to how to define such a system. System definition clearly depends upon the problem itself, but it also depends upon the parties whose interests might be affected by the resulting plan.

Reiner[1] refers to this dependence on the interests of many parties as a 'multiple goals framework.' In this framework modern planning problems face three basic problems: (1) how to define and arrange the complex system of multiple interests that would be affected by the plan so that it can be investigated effectively; (2) how to select (or develop) and apply analysis tools that can be used to analyze various planning options in this complex system; and (3) how to compare and contrast the options in the context of the goals of these multiple interests.

2.1.2. Energy system planning

Planning in energy systems is a special case of the general planning problem just outlined in that many energy planning problems depend primarily upon some underlying technology. Problems such as transcontinental pipeline development, large-scale energy research and development planning, and energy park development all have, as a base, some energy technology(ies).

The revolutionary change in energy problems in recent years has centered around the impact that a particular energy technology has on related economic or environmental problems, examples of which are the relatively recent societal concern over environmental quality or the Arab oil embargo (1973). This change has transformed many formerly traditional problems in energy into those of the modern variety just described. For instance, electric power plant siting was traditionally a very straightforward technical problem. Utilities sought sites that passed a number of technical tests such as availability of fuel and cooling water, suitable transportation access, and proximity to load centers. However, in recent years, environmental concern over air quality, thermal pollution and nuclear safety have significantly broadened the scope of consideration in power plant siting. As mentioned before, the energy park concept extends this scope even further to include socioeconomic and other factors.

1. See Reiner [15] or Davidoff and Reiner [4] who develop a 'choice theory of planning.'

This technology transformation experienced in power plant siting is typical of many current energy problems.[2] We can view this process as a kind of paradigm for energy planning. That is, the adoption of an energy alternative hinges upon satisfying a number of tests of feasibility which may be classified as the following: (a) scientific, (b) technological, (c) environmental, (d) economic, and (e) societal.

Scientific and technological feasibility imply the availability and applicability of scientific phenomena and their combination to accomplish a desired objective. The remaining tests of feasibility all serve as selectors of scientific and technologically viable alternatives.

Environmental feasibility addresses the impact that a technology and its scale of operation has on the quality of the environment. Should the environmental load be small enough so as not to degrade environmental quality the technology is environmentally feasible. Beneficial uses of environmental loads, a 'value added concept,' may, in some cases, enhance environmental quality.

Economic feasibility addresses cost of available technologies in accomplishing objectives. In general, the lowest cost technology is assigned the highest value of economic feasibility. However, secondary macroeconomic effects of a technology might be important as well. Both beneficial and detrimental secondary effects must be considered in determining overall economic feasibility, e.g. employment, GNP, financing, tax base, and so on.

Societal feasibility generally refers to the willingness of society to accept a technology. In particular, it involves the objectives of interests affected by the technology and the perceptions of those interests regarding the impacts of the technology. The larger the scale of the technology, the broader the scope of societal feasibility issues.

Viewing energy problems in the context of this technology transformation paradigm affords a simple explanation of many changes in energy patterns in recent years. For instance, again in the electric power industry, in the late 1960's and early 1970's we observed a marked shift in the northeastern United States from coal-fired to oil-fired electric power generation in response to societal concerns over environmental quality. A change in social priorities obligated this shift so that ambient air quality could be improved. In this case, both alternatives (oil and coal electric power generation) were technologically feasible but the relative societal acceptability of each prompted the shift from the *reference technology* (coal) to a *substitute* (oil). Implicit in the choice, however, was the assumption that

2. The original idea of technology transform analysis is attributed to Denton [5].

the other tests of feasibility were about equally satisfied by each alternative. Oil generation was cheap enough at that time to compete with the cost of coal generation and offered the advantage of lower sulphur dioxide emission as well; hence, it was selected by society.

Very few energy planning models at that time (around 1970) were equipped to deal with problems where a secondary objective, e.g. environmental quality maintenance, was elevated to equal or greater importance than that of the primary objective–economic efficiency.[3] This gap between energy problems and planning tools was widened in 1973 when unprecedented rises in oil prices changed the relative economically competitive position of oil. Indeed, the price of oil (economic feasibility) suggested a change back to coal-fired electric power generation from oil. But environmental pressures persist, the result of which is what is commonly referred to today as the 'energy-environment' dilemma.

In terms of the technology transformation paradigm, conventional coal and oil electric power plants were a scientific and technical reality in the sixties. They provided electric power economically and reliably enough to suit current standards and satisfied existing environmental criteria. Hence, until recently, the societal constraints which society places on performance (quantity of power delivered, reliability of service, etc) in the electric power industry could be regarded as quite rigid while the environmental and economic constraints were relatively much less rigid and, in fact, appeared minimal. Society demanded high levels of performance while fuel costs were sufficiently low to be of little concern and awareness of environmental problems was not widespread. However, recently imposed restrictions on sulphur dioxide emission and tough regulation of the electric power industry (prompted by increasing concern for environmental quality, the economic pressures associated with rising fuel prices, costs of pollution abatement equipment, and high capital costs) have caused the environmental and economic selectors of technology to become important enough not to be overlooked in energy planning.

As a result, since 1973 energy planning has emerged as a multiobjective problem, of the modern planning mold, where a variety of energy, economic and environmental tradeoffs must be examined. The energy modeling community has reacted in general only to the effects of the oil embargo (the 'Project Independence' ethic) by recognizing that economic forces are fundamental to current system dynamics. Their response has been a profu-

---

3. Prominent energy models of this period were developed by the National Petroleum Council [156], Dupree and West [128] and the U.S. Bureau of Mines [166].

sion of energy models that either impose or assume market clearing conditions on energy. [4] These models are normative in nature with cost minimization or profit maximization as their sole explicit norm. Some models have included resource and technological constraints while environmental impacts are often derived from economic and energy activities, i.e. treated as secondary output. [5] However, few of the current modeling efforts have attempted to relate the multiple objectives of various policy-making interests in an energy system to these models of energy, economic and environmental flows. [6]

2.2. ENERGY PLANNING AND MULTIPLE OBJECTIVES

We recall from the last section that there were three fundamental problems to be considered in modern planning: (1) the definition of the impacted system affected by the plan; (2) selection of appropriate analysis tools; and finally (3) analysis using these tools in a multiple goals framework.

Some insight into the first of these problems can be found in the theory of general systems. In this section some general systems concepts will be examined to see how they can contribute to defining and arranging complex systems.

2.2.1. General systems concepts

The literature abounds with definitions of general systems. However, discussions of systems in an abstract framework yield equally abstract definitions of a system and various terms associated with systems. These definitions must be much more rigorously specified in an applied framework where the results of analysis of a system, or some model of that system (which is itself a system), depend almost entirely upon the manner in which the system was first defined.

Perhaps it is best to first define a system in a broad sense [7] as a collection

---

4. These new models include Hudson and Jorgenson [139], Hoffman [138], Baughman [122], FEA PIES model [129] and the SRI-Gulf Energy Model [162].
5. Examples are Leontief [149], Herendeen [135], Just [146], and Strout [163].
6. Some of the few attempts at this include Mesarovic and Pestel [152] and Ford [131].
7. See Mihram [45] or general systems work of von Bertolantty [24], Klir [35], Laslo [39], Ashby [23], Ackoff and Emery [21], Sutherland [55], and others listed in the bibliography under the heading of systems theory.

of elements related so that it is reasonable to think of them collectively as a whole. In an abstract framework the relations of these elements to one or another can often be represented as functions. These functions form a model which is itself a system that approximates the system of interest.

The values of systems functions, which are often referred to as properties of an element of the system, vary with properties of other elements. Properties that are independent are termed system resources. The arrangement of functions and elements within a system then determine its structure. As a result, in the case of large complex systems, it may be very difficult, perhaps even impossible, to isolate the structure of the major subsystems to be studied.

In practice, three considerations are important in defining the limits for a system that is to be studied (see Kraemer [36]):

1. whether the limits encompass at least the area of activity where some knowledge of system relationships exists and can be readily determined;
2. whether these limits encompass all relevant elements that are regulated by important controlling interests in the system;
3. whether the limits delimit a system that can be effectively studied.

Often, simply by defining the functions that we wish or are able to consider, we implicitly define the limits of the system being studied. This collection of functions and the elements they relate is the system model.

Most models of systems are designed as operations on a set of system *inputs* (independent properties or variables regulated by interests in the system) which produce a set of system *outputs* (dependent variables). These outputs are defined as the state of the system to which one or more collections of inputs may produce.

An important question in general system theory is: 'How does one interpret the state of a system?' In application, this question becomes: 'Which functions should one choose to represent a good approximation of the system operation?' In general, this decision is made on the basis of the objectives or goals of the system. Hence, in practice, functions are used to represent relationships between elements determined by the goals and objectives of the system. Systems with goals and objectives are considered to be *'purposeful'* or teleological. [8]

---

8. The concept of 'purposeful system' is a central theme of the work of Ackoff and Emery [21].

In order to determine how well a particular state of the system fulfills the goals and objectives of the system, a performance measure is required. A major task in the analysis of physical systems is to devise ways to measure the system performance that do not affect this performance.

Results of analysis, i.e. comparison of various states of the system, can be used as feedback to alter the system goals and objectives and is generally the information sought in studying real systems. Systems that are able to alter performance on the basis of feedback are said to be *adaptive*.

We can also use the idea of system inputs and outputs to devise a more refined definition of a system (Mesarovic and Macko [43]):

A system S is a relation $S = X \cdot Y$ (cartesian product) on abstract sets X and Y. If the relationships between X (inputs) and Y (outputs) is a function, then S is a functional system, or a mapping $S:X \to Y$.

2.2.2. Systems and planning

We recall from the last section that the planning process is 'future directed'; hence, the ultimate aim of planning is to regulate the system being studied. This regulation is performed with respect to the goals and objectives used in measuring the state of the system.

The problem with applying these systems concepts to modern planning problems is that problems are posed in a multiple goals framework. The multiple goals are objectives of controlling interests in the system being studied. For example, returning to the energy park problem, the development of an energy park in a region would be influenced by a wide variety of these interests. We might refer to these interests as actors and the overall system that would be affected by the energy park development as a planning system.

We see that in modern planning systems of this sort there may be no overall objectives in the system but rather a collection of individual actor objectives, all of which must be considered. The problem then is to structure or arrange this collection of actor interests in a manner that will reveal relationships with the operation of the system being studied.

It is convenient to think of the adoption of a particular collection of individual actor objectives as a policy; hence, the actor might be referred to as a *policy-making interest* in the planning system. Actors in a planning system make decisions based upon the decisions of other actors as well as in response to the current state of the system. Hence, the system is *adaptive* in the sense discussed earlier.

2.2.3. Energy planning systems

We recall that the principal purpose of energy planning is to formulate policy. However, we found in the last section that the overall policy must address the variety of policy-making interests in modern planning systems.

Perhaps the most powerful way to arrange a collection of policy-making interests so that it can be studied as a unit is to use a hierarchical structure. A hierarchy originally denoted a vertical authority structure in human organizations (Herbert Simon in Pattee [47]). In the theory of complex systems [9] 'hierarchy' has taken a more general meaning. Simon (Pattee [46]) describes a hierarchy as a 'set of Chinese boxes that are constructed to sequentially enclose each other for as long as the patience of the craftsman holds out.' This Chinese box hierarchy is a sequence of complete ordering whereas some hierarchies, such as the original human organizations concept, are partial orderings or trees that represent some order of dominance.

This dominance hierarchy is of interest in structuring the policy-making interests in an energy planning system since we wish somehow to derive a set of collective objectives for this array of individual policy-making interests. The collective interests of this array form the top level of the hierarchy while the individual objectives of each policy maker form lower levels of the hierarchy. In addition, we wish to capture the relative influence that a policy maker can bring to bear in fulfilling his objectives. The manner in which the objectives of a policy-making interest vary in itself forms a system. We might refer to this system as a *purposeful subsystem* of the original system. [10]

We shall see in what follows how these basic hierarchical concepts can be used to structure the objectives and influence of policy-making interests in a planning system. We shall call the resulting hierarchically arranged system a *policy-making system.*

In a later section the special considerations that are necessary in dealing with energy planning systems will be reviewed. It will turn out that energy planning systems can be treated as a special case of more general concept of policy-making systems.

We now consider a formal description of policy-making systems. We shall be able to operationalize some elements of this formalism in structuring energy planning systems.

9. Some general references on the study of hierarchies and Pattee [47], Mesarovic and Macko [43], Whyte et al. [58] and Weiss [57].
10. A purposeful system in the sense of Ackoff and Emery [21] and Emery [32].

2.3. STRUCTURE OF POLICY-MAKING SYSTEMS

Some systems have no general overall purpose but rather have purposes for each of their subsystems. The interaction of these subsystems with their individual purposes drives the overall system from state to state. Subsystems are structured among themselves hierarchically so that some of them on higher levels control and regulate others on lower levels. Regulation is the control of subsystems to adjust their functions in harmony with those of other subsystems in order to produce overall system stability. To make regulation feasible, feedback among subsystems is used. Feedback means modification of subsystems' output by actually modifying the input in order to improve the stability of the entire system. This is an important element in using the results of analysis in the planning process.

Let us formalize the structure of interactions of such a complex system. A *policy-making system* consists of:

1. a set of policy makers which form purposeful subsystems of the system considered;
2. a set of variables used to define the objectives and constraints of the policy makers; and
3. a set of relationships between the policy makers and the state variables.

The state of the policy-making systems at time t is described in terms of *system variables*. $X(t)$ and $Y(t)$ defined as follows:

$X(t)$ = a vector which consists of all *decision* or *control variables*, the values of which are determined by policy makers in period t;

$Y(t)$ = a vector which defines the *state* of the physical system at the end of period t (*state variables*);

$Z(t)$ = the *history* of the system up to period t, and is defined by:

$Z(t) = \Pi_{s=1}^{t}\{X(s),Y(s)\}$ which is the cartesian product of previous values of X and Y. The current state of the policy-making systems is related to $Z(t-1)$ by the following *system process function*:

$$Y(t) = F\{t, X(t), Z(t-1)\} \qquad (2.3.1)$$

with the system constraints: $H\{t, X(t), Z(t-1)\} \leqq b(t)$

As we shall see later, in a complex system, $Y(t)$ is hierarchically structured. The policy makers themselves in the policy-making system are also hierarchically structured. Included in equation (2.3.1) are accounting identities (such as the level of resources still available, or the profits of an industry) and restrictions on the system such as resource and technological constraints. The interactions among policy-making subsystems might be governed by a set of institutional rules and procedures.[11]

Suppose that we have n policy makers, P_i, $i = 1, 2, \ldots, n$. We define:

1. $X_i(t)$ = the decision variables controlled by policy maker P_i during period t.
2. $\tilde{X}_i(t)$ = future decision variables of P_i; $\tilde{X}_i(t) = \Pi_{s=t+1}^{\infty} X_i(s)$ (a Cartesian product).
3. $Y_i(t)$ and $\tilde{Y}_i(t)$ are the current and future system variables of concern to P_i.
4. b_i and $\tilde{b}_i$ are current and future decision constraints set by P_i.

We next define the notion of *monitor* or *policy variables*:

5. $V_i(t)$ = current system variables important to fulfilling the objectives of P_i. $V_i(t)$ is equal to some mathematical projection of the history of the system $Z(t)$.
6. $\tilde{V}_i(t)$ = future state variables important to fulfill his objectives in period t. $\tilde{V}_i(t)$ is equal to some mathematical projection of $\Pi_{s=t+1}^{\infty}\{X(s), Y(s)\}$.

We shall represent the pairs $\{X_i(t), \tilde{X}_i(t)\}$, $\{Y_i(t), \tilde{Y}_i(t)\}$, $\{V_i(t), \tilde{V}_i(t)\}$ and $\{b_i(t), \tilde{b}_i(t)\}$ by $\overline{X}_i(t)$, $\overline{Y}_i(t)$, $\overline{V}_i(t)$ and $\overline{b}_i(t)$, respectively. A policy is defined by the choice of $\overline{V}_i(t)$ and by its rate of change.

Each P_i has a set of objectives $g_i(t)$ that are functions of $\overline{V}_i(t)$ and may change over time (see for example Shakun-Lewin's theory of situational normativism – Lewin and Shakun [41]). Thus, how well a subsystem P_i attains its objectives is related not only to its history of the system $Z_i(t)$ but also to $\tilde{V}_i(t)$, the forecasted future state of the system. In addition, the decisions of other policy makers $\overline{X}_j(t)$, $j \neq i$, are related to the objectives of P_j.

Policy maker P_i has a perception (or model) of how his and other policy makers' decisions would affect the future of the system and, hence, his policy variables $\overline{V}_i(t)$. Thus, there is a perceived process function f_i such

---

11. For more discussion on institutional rules and procedures see Saaty, Ma and Blair [160].

that:

$$\mathbf{Y}_i(t) = f_i\{t, \overline{\mathbf{X}}(t), \mathbf{Z}(t - 1)\} \tag{2.3.2}$$

and subsystem perceived constraints:

$$h_i\{t, \overline{\mathbf{X}}(t)\} \leqq \overline{b}_i(t) \tag{2.3.3}$$

Although the forms of the functions f_i and h_i are not usually known, this general formulation serves to characterize how purposeful subsystems interact.

In a policy-making system, decision makers will attempt to influence, bargain or negotiate with other decision makers to achieve their objectives. Thus, in addition to direct control of his own $\mathbf{X}_i(t)$ and $b_i(t)$, a policy maker P_i has various degrees of indirect influence on $\mathbf{X}_j(t)$ and $b_j(t)$, where $j \neq i$, through a set of institutional rules and procedures. Also P_i may have to compromise his multidimensional objectives g_i (a vector-valued function) during the negotiation process. To represent this interactive decision process mathematically for each P_i, $i = 1, 2, \ldots, n$, we have the following dynamic optimization problem:

Find $\overline{\mathbf{X}}_i(t)$ and $\overline{\mathbf{X}}_j(t)$, $j \neq i$, which:

maximizes $g_i\{t, \overline{\mathbf{V}}_i(t)\}$ subject to the process functions and constraints:

$$\tag{2.3.4}$$

$$\mathbf{Y}_i(t) = f_i\{t, \overline{\mathbf{X}}(t), \mathbf{Z}(t - 1)\} \tag{2.3.5}$$

$$h_i\{t, \overline{\mathbf{X}}(t), \mathbf{Z}(t - 1)\} \leqq \overline{b}(t) \tag{2.3.6}$$

and also subject to a set of structural constraints known as system interaction rules and procedures such as, for example, a utility rate hearing.

The latter constraints are complex and are not stated analytically here but will eventually need formalization. Note that $\mathbf{X}_i(t)$ represents a commitment by P_i concerning his future actions.

Therefore, in order for a policy-making system to evolve from one dynamically stable state $\{\mathbf{X}(t), \mathbf{Y}(t)\}$ to another such state $\{\mathbf{X}(t + 1), \mathbf{Y}(t + 1)\}$, there are two requirements that must be met: (1) all of P_i must arrive at a set of decisions which satisfy the 'perceived' constraints in equations (2.3.4), (2.3.5) and (2.3.6) and they should have arrived at a prescribed level for their objectives; and (2) their decisions must satisfy the overall system constraints in (1). The formalism for policy-making systems is summarized in table 2.1.

In the next section we consider a special case of the general policy-making system. This special case is applied to energy planning. By borrowing from the theory of Leontief's input-output model we will develop a specific model of the system process functions for energy systems that were only abstractly defined for the general case of policy-making systems.

2.4. ENERGY-ENVIRONMENT SYSTEMS

In energy planning we are often concerned with a special kind of policy-making system. This system, often called an *energy-environment system*, is a collection of primarily economic activities that consume energy and affect environmental quality. The activities are industries producing goods and services. An industry, in the process of producing its product, consumes varying amounts of one or more energy types (coal, crude oil, natural gas, etc) and generates a number of pollutants (SO_2, NO_x, hydro-

TABLE 2.1. Policy-making systems formalism.

Time	Variables and functions	Policy-making system	Purposeful subsystems	Exogenous variables
Current and future	Decision variables	$X(t)$	$X_i(t), \check{X}_i(t)$	$X_0(t)$
	State variables	$Y(t)$	$Y_i(t), \check{Y}_i(t)$	
	Policy variables		$V_i(t), \check{V}_i(t)$	
	Objective variables		$g_i\{t, \check{V}_i(t)\}$	
	Constraint variables	$b(t)$	$b_i(t), \check{b}_i(t)$	$b_0(t)$
	Process function	$Y(t) = F\{t, X(t), Z(t-1)\}$	$\overline{Y}(t) = f_i\{t, \overline{X}(t), Z(t-1)\}$	
	Constraint function	$H\{t, X(t), Z(t-1)\} \leq b$	$h_i\{t, \overline{X}(t), Z(t-1)\} \leq \overline{b}_i$	
Past	History	$Z(t)$		

The Dynamic Optimization Problem:
  Find $\overline{X}_i(t)$ and $\overline{X}_j(t)$, $j \neq i$, which maximizes $g_i\{t, \nabla_i(t)\}$ subject to the process functions and constraints:

$$\overline{Y}_i(t) = f_i\{t, \overline{X}(t), Z(t-1)\}$$
$$h_i\{t, \overline{X}(t), Z(t-1)\} \leq \overline{b}_i(t)$$

and system interaction rules and procedures.

carbons, waste heat, etc). Each industry also requires the outputs of other industries that become inputs to its own industrial process. Thus, the industrial processes of all these industries, the energy they consume and the pollution they generate are interdependent. Such interdependent processes were abstractly defined in the last section as system process functions.

Most current models of energy-environment systems have been detailed analytical descriptions of systems process functions that were constructed without considering the policy makers that are affected by these functions. The policy makers in an energy-environment system are those interests that control and whose objectives are affected by the industrial activities. Consequently, there is a need for relating the individual objectives of policy makers to these system process functions.

In order to provide this relation, an analytical representation of the system process functions must be adopted so that each can be explicitly formulated as a function of factors that the policy makers control. Since activities in energy-environment systems are primarily economic a convenient representation of these functions can be developed in the economist's input-output framework. The following discussion will describe how the process functions in an energy-environment system can be constructed from the basic concepts of input-output.

Input-output in economics. In basic economic input-output theory,[12] all information is derviced from an interindustry transactions matrix. The rows of such a matrix describe the distribution of a producer's output measured in dollars throughout other industries in the economy. A column describes the composition of the inputs required by an industry to produce its output (see table 2.2).

The most commonly used approach to input-output (I/O) analysis is the 'open' model which includes the interindustry transactions just described and the following rows and columns which are appended to the interindustry transactions matrix:

1. *Sales to final demand.* Some additional columns (table 2.2) are included depicting the sales of a producing industry to final markets (sales to households, government, exports, etc). The sum of all sales to final markets for all industries is the gross product of an economy.

---

12. Modern economic input-output analysis is all developed from the original work of Leontief [149].

TABLE 2.2. Input-output flow table.

		Agriculture	Mining	Construction	Manufacturing	Trade	Transportation	Services	Other	FINAL MARKETS — Persons	Investors	Foreigners	Government
PRODUCERS	Agriculture									Personal consumption expenditures	Gross private domestic investment	Net exports of goods and services	Government purchases of goods and services
	Mining												
	Construction												
	Manufacturing												
	Trade												
	Transportation												
	Services												
	Other												
VALUE ADDED	Employees	Employee compensation											
	Owners of Business and Capital	Profit-type income and capital consumption allowances											
	Government	Indirect business taxes											

GROSS NATIONAL PRODUCT

Source: U.S. Department of Commerce, Bureau of Economic Analysis.

2. *Value added by manufacturer.* Some additional rows are also included that depict the incomes generated by a producer's production process or the value added by his enterprise (employee compensation, profits, taxes, etc). The sum of all value-added items for all industries also yields the gross regional product.

If we sum across the columns of the interindustry transactions matrix, i.e. add up all of a producer's sales to intermediate consumers (all except final consumers), we obtain the industry's intermediate output. If we then add the producer's sales to final demand to his intermediate output, the result is the total gross output. Similarly if we sum down the rows of the transactions matrix, i.e. add up all of a producer's purchases from other industries, we obtain the producer's intermediate input. The value of this intermediate input plus the value added by his industrial process yields the total industry inputs. By convention, total outputs in an input-output economy are set equal to total inputs.

Of fundamental interest in economic I/O analysis is the matrix of *technical coefficients* or direct requirements. This matrix is the result of taking each column of the interindustry transactions matrix and dividing each element of a particular column by the corresponding total output of the industry whose inputs the column represents. The result is the dollars worth of input of each industry in the economy that is required to produce one dollar's worth of a producer's output. Let us adopt the following notation: Z = matrix of interindustry transactions; Y = vector of total sales to final demands; and X = vector of total industry outputs. Z is an $n \times n$ matrix where n is the number of industries in the economy. Y and X are column vectors of dimension $n \times 1$. An element of the technical coefficients matrix is given by:

$$a_{ij} = z_{ij}/x_j \qquad\qquad \begin{array}{l} i = 1, 2, \ldots, n \text{ producers} \\ j = 1, 2, \ldots, n \text{ consumers} \end{array}$$

or in matrix notation:

$$A = Z\hat{X}^{-1}, \text{ where } \hat{X}^{-1} = \begin{bmatrix} 1/x_1 & & & & 0 \\ & 1/x_2 & & & \\ & & \cdot & & \\ & & & \cdot & \\ & & & & \cdot \\ 0 & & & & 1/x_n \end{bmatrix}$$

We recall that the sum of intermediate outputs, which is given by the product $\mathbf{AX}$ and final demands yields total outputs:

$$\mathbf{AX} + \mathbf{Y} = \mathbf{X}.$$

Hence, it follows by ordinary matrix algebra:

$$\mathbf{X} - \mathbf{AX} = \mathbf{Y}$$
$$(\mathbf{I} - \mathbf{A})\mathbf{X} = \mathbf{Y}$$
$$\mathbf{X} = (\mathbf{I} - \mathbf{A})^{-1}\mathbf{Y}. \qquad (2.4.1)$$

Equation (2.4.1) is the fundamental equation of input-output analysis as developed by Leontief. The matrix $(\mathbf{I} - \mathbf{A})^{-1}$ is often referred to as the Leontief inverse or the total requirements matrix since it reflects the total inputs required (directly to produce an industry's product and indirectly to produce that industry's inputs) of all industries to produce one dollar's worth of the industries' output. $(\mathbf{I} - \mathbf{A})^{-1}$ is the transformation from final products to total outputs; similarly, $(\mathbf{I} - \mathbf{A})$ provides the transformation from total outputs to final products.

Generalized input-output analysis. Input-Output has been extended in recent years to include consideration of other factors such as energy consumption, pollution emission and employment that are directly related to inter-industry activity.[13] These extensions can be used to represent a set of system process functions in an energy-environment system. In general, these extensions provide measures of energy consumption, pollution and employment that vary linearly with industry output. Hence, total amounts of these quantities can be computed from the value of the industrial output by multiplying by a set of coefficients (e.g. tons of SO_2 generated per dollar of output or tons of coal consumed per dollar of output).

Let u_{ik} be the amount of energy of type k that is consumed per dollar of output of industry i, so that:

$$u_{ik}^* = u_{ik}x_i, \text{ where } u_{ik}^* \text{ is the total energy of type } k \text{ consumed by}$$
industry i and x_i is the total output of industry i.

If we have K energy types in the economy, then we can construct a matrix

13. See for example Herendeen [134], White [170], Leontief [148], Folk and Hannon [130], Victor [169], Miernyk and Sears [153].

u, a row of which is $u_{1k}, u_{2k}, \ldots, u_{nk}$, for $k = 1, 2, \ldots, K$ rows. The product
uX gives a vector, the elements of which are the total energy consumed by
the economy of each energy type $\mathbf{u}^*$ (more comprehensive formulations
also include the energy consumed by final demand)[14].

$$\mathbf{u}^* = \mathbf{uX}$$

A method for constructing this matrix **u** in practice, with available data, is
described in chapter 3 and Appendix A.

One can construct coefficient matrices similar to **u** for pollution emission
and employment:

> $\mathbf{v}^* = \mathbf{vX}$, where an element of **v** is the amount of a particular pol-
> lutant produced per dollar of industry output. The vector
> $\mathbf{v}^*$ gives the total emission of each pollutant generated in
> the economy.
>
> $\mathbf{w}^* = \mathbf{wX}$, where **w** is of dimension $1 \times n$ and gives the number of
> employees in an industry per dollar of output. The scalar
> w^* is the total employment of the economy.

By concatenating the matrices of coefficients (**u**, **v**, **w**) we can construct a
total intensity matrix **D**:

$$\mathbf{D} = \begin{bmatrix} \mathbf{u} \\ \hline \mathbf{v} \\ \hline \mathbf{w} \end{bmatrix}$$

so that; $\mathbf{X}^* = \mathbf{DX}$, $\mathbf{X}^* = \begin{bmatrix} \mathbf{u}^* \\ \hline \mathbf{v}^* \\ \hline w^* \end{bmatrix}$

A complete model of the system process functions in an energy environment system can then be given by Λ and **D**:

$$\mathbf{X} = (\mathbf{I} - \mathbf{A})^{-1} \mathbf{Y}$$
$$\mathbf{X}^* = \mathbf{DX}$$

---

14. These formulations usually include energy-final demand coefficients, e.g. $\mathbf{u}^* = \mathbf{u}$
$(\mathbf{I} - \mathbf{A})^{-1} + \mathbf{S}$, where **S** has units of Btus per dollar of final demand. This formulation
is discussed in more detail in Appendix A.

or more simply: $\mathbf{X}^* = \mathbf{D}^*\mathbf{Y}$, where $\mathbf{D}^* = \mathbf{D}(\mathbf{I} - \mathbf{A})^{-1}$. $\mathbf{X}^*$ (the levels of energy consumption, pollution and employment) constitutes a set of *state variables* whose values are determined by the values of $\mathbf{Y}$ (a set of *decision variables*) by means of the functional relationship provided by $\mathbf{D}^*$. Note that if we wish to include the values of industry outputs in the set of state variables we might represent the model of system process functions by the following:

$$\mathcal{X} = \mathbf{T}^*\mathbf{Y} \qquad \mathbf{T}^* = \left[\frac{(\mathbf{I} - \mathbf{A})^{-1}}{\mathbf{D}^*} \right]$$

$$\mathcal{X} = (\mathbf{X}, \mathbf{X}^*)$$

2.5. THE EIGENVALUE PRIORITIZATION MODEL

> 'Objectives and their order of priority are determined largely by value judgments rather than by analysis. These judgments should be made explicitly'.
>
> (Kahn [10] in his investigation of social planning).

The method derived from Saaty's theory centers around constructing a matrix of pairwise comparisons of activities. The entries of this matrix indicate the dominance of one activity over another with respect to a specific comparison criterion. These entries are made on a ratio scale, i.e. an entry a_{ij} is taken as an estimate of the ratio w_i/w_j, the dominance of ith compared to the jth activity.

This matrix scaling formulation translates to a largest eigenvalue problem. The Perron-Frobenius[15] theory of positive matrices ensures that a unique positive eigenvalue exists for matrices with only positive entries. The corresponding normalized eigenvector is the vector of relative weights of the activities being compared.

In applying the method one is typically faced with a comparison of n objects in pairs according to their relative weights (dominance). Hence,

---

15. The Perron-Frobenius work is summarized in Wielandt [59] Gantmacher [33] and Householder [34].

we can represent these pairwise comparisons as a matrix:

$$
\mathbf{A} =
\begin{bmatrix}
w_1/w_1 & w_1/w_2 \ldots & w_1/w_n \\
w_2/w_1 & w_2/w_2 \ldots & w_2/w_n \\
\cdot & \cdot & \cdot \\
\cdot & \cdot & \cdot \\
\cdot & \cdot & \cdot \\
w_n/w_1 & w_n/w_2 \ldots & w_n/w_n
\end{bmatrix}
$$

As an example consider n stones whose normalized weights are given by the vector $\mathbf{w} = (w_1, w_2, \ldots, w_n)$, $(\Sigma_i^n w_i = 1)$. Suppose we know only the ratios of the weights of the stones, i.e. $a_{ij} = w_i/w_j$ is the ratio of the weight of stone i to that of stone j. Hence, we form the matrix of these ratios (pairwise comparisons). We assume that this matrix has positive entries everywhere and satisfies the reciprocal property $a_{ji} = 1/a_{ij}$. Such a matrix is called a reciprocal matrix.

It is clear that if we knew the absolute weights of the stones (normalized by the total weight of all the stones), $\mathbf{w} = (w_1, w_2, \ldots w_n)$ then the following relation would hold:

$$
\mathbf{A}\mathbf{w} = n\mathbf{w} =
\begin{bmatrix}
w_1/w_1 & w_1/w_2 \ldots & w_1/w_n \\
w_2/w_1 & w_2/w_2 \ldots & w_2/w_n \\
\cdot & & \cdot \\
\cdot & & \cdot \\
\cdot & & \cdot \\
w_n/w_1 & w_n/w_2 \ldots & w_n/w_n
\end{bmatrix}
\begin{bmatrix}
w_1 \\
w_2 \\
\cdot \\
\cdot \\
\cdot \\
w_n
\end{bmatrix}
=
\begin{bmatrix}
nw_1 \\
nw_2 \\
\cdot \\
\cdot \\
\cdot \\
nw_n
\end{bmatrix}
\tag{2.5.1}
$$

However, if we did not know the weights of the stones and had only the matrix of pairwise comparisons $\mathbf{A}$, we would have to solve equation (2.5.1) for $\mathbf{w}$. In a somewhat more familiar form this equation becomes:

$$
(\mathbf{A} - n\mathbf{I})\mathbf{w} = 0 \tag{2.5.2}
$$

We can easily recognize this as an eigenvalue problem, i.e. a problem for which a non-zero solution exists if and only if n is an eigenvalue of $\mathbf{A}$. Two results in the theory of positive matrices are important in dealing with this problem: (1) the matrix $\mathbf{A}$ is of unit rank since every row is a constant multiple of the first row – as a result all the eigenvalues of $\mathbf{A}$, λ_i, $i = 1$, $2, \ldots, n$ are zero except one; and (2) it is well known that the sum of the eigenvalues of a positive matrix is equal to the trace of that matrix (the sum of the diagonal elements). As a result, since $a_{ii} = 1$ in reciprocal matrices:

$$\sum_{i=1}^{n} \lambda_i = \text{tr} (\mathbf{A}) = n$$

For reciprocal matrices, as a result of (1) and (2), only one of the λ_i (which we refer to as λ_{max}) is equal to n and $\lambda_i = 0$ for $\lambda_i \neq \lambda_{max}$. Similarly, $\mathbf{w}$ is any normalized column of $\mathbf{A}$. Hence, by solving the eigenvalue problem we retrieve the vector of weights from the matrix of pairwise comparison.

We can also note that in our matrix $\mathbf{A}$, in the case of the stones, the pairwise comparisons are completely consistent with one another, i.e. $a_{ij} a_{jk} = a_{ik}$ (the 'strong' consistency property – transitivity). The reciprocal property mentioned earlier is referred to as the 'weak' consistency property $a_{ji} = 1/a_{ij}$. If a matrix is strongly consistent we can easily construct the entire matrix from a single row, e.g. the first row:

$$a_{jk} = a_{lk}/a_{1j} \quad (a_{1j} \neq 0)$$

The most important notions of Saaty's theory are useful in dealing with problems where the underlying scale ($\mathbf{w}$) is not known but where we do have estimates of the entries of $\mathbf{A}$. We do not require the strong consistency property to hold. This realistically allows for inconsistency in judgements, since despite people's best efforts their pairwise comparisons of activities are often to some degree inconsistent and intransitive.[16] This inconsistent case (actually only weakly consistent) is of interest in deriving an underlying scale. For any matrix, small perturbations in the entries imply similar perturbations in the eigenvalues.[17] The eigenvalue problem for the inconsistent case then becomes:[18]

$$\mathbf{A}\mathbf{w}' = \lambda_{max} \mathbf{w}'$$

The closer λ_{max} is to n, the more consistent the judgements implicit in $\mathbf{A}$. Saaty [49] establishes the index $\lambda_{max} - n/n - 1$ as a measure of overall consistency of judgements.[19] He proves that $\lambda_{max} \geqq n$ always, for reciprocal matrices. Actually, this result is only a corollary to a theorem that states

16. Saaty [49] discusses this notion of judgemental consistency in some detail.
17. Perturbations of eigenvalues are discussed in Wilkinson [60].
18. The non-principal eigenvalues will be non-zero in the inconsistent case but will tend to zero as λ max tends to n.
19. The index $\alpha = (\lambda - n)/(n - 1)$ is really the average of the non-principal eigenvalues.

that a pairwise comparison matrix $\mathbf{A}$ is consistent if and only if $\lambda_{max} = n$. If $\lambda_{max} \neq n$ then the other eigenvalues will be non-zero, perhaps complex.

Saaty discusses at length the question of the appropriate scale to be used in the pairwise comparison process. He argues that the scale should satisfy three fundamental requirements:

1. It should be possible to represent decision makers' differences in making comparisons. It should represent as much as possible all distinct shades of choice the decision makers may express.
2. If we denote the scale values by $x_1, x_2, \ldots, x_p$ then:

$$x_{i+1} - x_i = 1 \qquad\qquad\qquad i = 1, 2, \ldots, p - 1$$

3. The decision maker must be aware of all gradations of the scale at the same time.

The third requirement is particularly important. Miller's [46] psychological experiments show that a decision maker cannot simultaneously compare more than seven (± 2) objects without confusion. With a unit difference between values (Requirement 2 above) and the assumption that $x_1 = 1$ is the identity comparison, a reasonable scale would range from 1 to 9.

In using the scale for pairwise comparison Saaty proposes the ranks in the scale as described in table 2.3. The values used in pairwise comparison are nearest integer approximations of relative dominance. Saaty [49] presents a number of statistical experiments, the results of which support the resultant comparison scale.

Hierarchical application of the eigenvalue method entails assigning priorities (weights) to activities according to their impact on higher level objectives which are, in turn, evaluated in terms of still higher objectives. Normalized eigenvectors corresponding to dominant eigenvalues of lower levels form the columns of matrix which is multiplied by the dominant eigenvector of the next higher level. The influence at a lower level is transferred in this manner to higher levels.

We now consider a simple example of hierarchical application of the eigenvalue model which might serve better to illustrate the approach than a formal description of the theory which can be found in Saaty [49]. Later, in chapter 3, we will return to this example when the hierarchical model is modified to link the objectives and influence of policy makers with relevant system variables in an energy-environment system.

TABLE 2.3. Eigenvalue prioritization model: judgement scale.

Intensity of importance	Definition	Explanation
1	Equal importance	Two activities contribute equally to the objective
3	Weak imprtance of one over another	Experience and judgement slightly favor one activity over another
5	Essential or strong importance	Experience and judgement strongly favor one activity over another
7	Demonstrated importance	An activity is strongly favored and its dominance is demonstrated in practice.
9	Absolute importance	The evidence favoring one activity over another is of the highest possible order of affirmation
2, 4, 6, 8	Intermediate values between the two adjacent judgements	When compromise is needed
Reciprocals of above non-zero	If activity *i* has one of the above non-zero numbers assigned to it when compared with activity *j*, then *j* has the reciprocal value when compared with *i*	
Rationals	Ratios arising from the scale	If consistency were to be forced by obtaining *n* numerical values to span the matrix

Source: Saaty [49].

A hierarchical energy-environment system example.[20] Consider three actors in a simple energy environment system:

1. the *government*, which regulates the consumption of energy, the maintenance of environmental quality and influences the level of reserves of energy resources;
2. a private profit motivated *utility* which generates electricity at the rate

---

20. This example is taken from P. Blair, 'Working Paper on Phase I Methodology,' included in ERDA Progress Report COO-2547-2, 'Planning for a Program for Regional Energy Analysis,' March 31, 1975.

demanded by a consuming public and which is subject to the regulatory influence of the government;
3. *consumers* of electrical energy who are concerned about the price of energy, as well as the quality of the environment in which they live.

The actors can pursue any or all of the following objectives, some of which will be pursued more vigorously than others:

1. minimizing the cost of producing and consequently consuming energy;
2. maintenance of a high standard of environmental quality;
3. preservation or conservation of precious natural energy resources.

Let us presume that the factors or variables that measure the state of the energy-environment system are the following:

1. the cost of energy in dollars;
2. the quality of the environment in pollution units (e.g. concentration of SO_2);
3. the proven reserves of natural energy resources.

Suppose now that the above energy-environment system has three alternatives for producing electric power in the future. These three alternatives constitute the possible future scenarios of the energy-environment system:

1. use of oil-fired plants;
2. Use of coal-fired plants;
3. use of nuclear plants.

Given this information, how can we determine the most desirable future of the energy-environment system based on the interaction of the defined actors who are pursuing individual objectives? In the terms of the eigenvalue approach we can first construct a matrix of pairwise comparisons of a particular actor's influence (effectiveness in realizing his goals) as compared to the relative influence of other actors. We recall from the last section (table 2.3) that the eigenvalue model requires that the judgements be based on a scale from zero to nine, with the inverse relationship defined by the reciprocal.

In this case, for instance, the government may exert more influence on the state of the energy-environment system as compared to the utility by, say, a factor of 3. Accordingly, the influence of the utility as compared to the government is $\frac{1}{4}$. An example of these comparisons for all actors constructed in a matrix might be the following:

$$
\begin{array}{c}
\quad\; A_1 \quad A_2 \quad A_3 \\
\begin{array}{c} A_1 \\ A_2 \\ A_3 \end{array}
\begin{bmatrix}
1.00 & 3.00 & 2.00 \\
.33 & 1.00 & 4.00 \\
.50 & .25 & 1.00
\end{bmatrix}
\end{array}
\qquad
\begin{array}{l}
A_1 = \text{government} \\
A_2 = \text{utility} \\
A_3 = \text{consumers}
\end{array}
$$

The somewhat naive notion of this 'overall' influence will be refined later to a measure of influence over important system variables in an energy-environment system.

According to the theory of the eigenvalue method the normalized eigenvector corresponding to the dominant eigenvalue of this matrix can be used as a measure of dominance of one actor compared to another with respect to relative influence on the energy environment system. Let us call this matrix $\underline{W}$, the dominant eigenvector of which is A.

$$
\bar{A} = \begin{bmatrix} 0.532 \\ 0.322 \\ 0.146 \end{bmatrix}
\begin{array}{l} \text{government} \\ \text{utility} \\ \text{consumer} \end{array}
$$

$\bar{A}$ is then the measure of the relative influence of each actor. Each actor pursues the defined objectives with varying degrees of emphasis. Pairwise comparison of the defined objectives of each actor can accordingly be made to determine the priority with which an actor views each of the sets of objectives.

Let us call the matrix of comparisons of objectives for the government actor A_1, that for the utility A_2, and that for the consumer A_3. A typical set of matrix constructions of these comparisons might be:

$$
\begin{array}{cc}
\begin{array}{c}
A_1 = \text{government:} \\
\quad\; O_1 \quad O_2 \quad O_3 \\
\begin{array}{c} O_1 \\ O_2 \\ O_3 \end{array}
\begin{bmatrix}
1.00 & 3.00 & 3.00 \\
0.33 & 1.00 & 0.33 \\
0.33 & 3.00 & 1.00
\end{bmatrix}
\end{array}
&
\begin{array}{c}
A_2 = \text{utility:} \\
\quad\; O_1 \quad O_2 \quad O_3 \\
\begin{array}{c} O_1 \\ O_2 \\ O_3 \end{array}
\begin{bmatrix}
1.00 & 8.00 & 4.00 \\
0.13 & 1.00 & 0.17 \\
0.25 & 6.00 & 1.00
\end{bmatrix}
\end{array}
\end{array}
$$

A_3 = consumer:

$$
\begin{array}{c}
\quad\; O_1 \quad\; O_2 \quad\; O_3 \\
\begin{array}{c} O_1 \\ O_2 \\ O_3 \end{array}
\begin{bmatrix}
1.00 & 2.00 & 3.00 \\
0.50 & 1.00 & 2.00 \\
0.33 & 0.50 & 1.00
\end{bmatrix}
\end{array}
$$

O_1 = minimizing energy cost
O_2 = maintaining environmental
      quality
O_3 = preserving natural resources

The dominant eigenvectors of A_1, A_2 and A_3, which we can call $\bar{g}_1$, $\bar{g}_2$ and $\bar{g}_3$, respectively, describe the relative priorities of the objectives for each actor:

$$
\bar{g}_1 = \begin{bmatrix} 0.584 \\ 0.135 \\ 0.281 \end{bmatrix} \begin{array}{c} O_1 \\ O_2 \\ O_3 \end{array}
\qquad
\bar{g}_2 = \begin{bmatrix} 0.691 \\ 0.060 \\ 0.249 \end{bmatrix} \begin{array}{c} O_1 \\ O_2 \\ O_3 \end{array}
\qquad
\bar{g}_3 = \begin{bmatrix} 0.540 \\ 0.297 \\ 0.163 \end{bmatrix} \begin{array}{c} O_1 \\ O_2 \\ O_3 \end{array}
$$

Concatenation of these vectors in a matrix G and multiplication by $\bar{A}$ yields a vector $\bar{B}$ which is a weighting of objectives based on the relative importance of each objective to each actor and the relative influence of each actor:

$$
\bar{B} = G\bar{A} = \begin{bmatrix}
0.584 & 0.691 & 0.540 \\
0.135 & 0.060 & 0.297 \\
0.281 & 0.249 & 0.163
\end{bmatrix}
\begin{bmatrix}
0.532 \\
0.322 \\
0.146
\end{bmatrix}
= \begin{bmatrix}
0.612 \\
0.135 \\
0.253
\end{bmatrix}
\begin{array}{c} O_1 \\ O_2 \\ O_3 \end{array}
$$

We can again apply the scheme in the next hierarchy to determine the importance of a particular factor in pursuing an objective. This requires pairwise comparison again, this time of the three factors with respect to each objective. Let us call the comparison matrices for objective 1 (minimizing energy cost) O_1; for objective 2 (maintaining environmental quality) O_2; and for objective 3 (preserving natural resources) O_3, and their respective dominant eigenvectors $\bar{h}_1$, $\bar{h}_2$, $\bar{h}_3$ (concatenated to yield H):

O_1 = minimize
    energy cost

$$
\begin{array}{c}
\quad\; f_1 \quad\; f_2 \quad\; f_3 \\
\begin{array}{c} f_1 \\ f_2 \\ f_3 \end{array}
\begin{bmatrix}
1.00 & 8.00 & 3.00 \\
0.13 & 1.00 & 0.14 \\
0.33 & 7.00 & 1.00
\end{bmatrix}
\end{array}
$$

O_2 = maintain quality of
    environment

$$
\begin{array}{c}
\quad\; f_1 \quad\; f_2 \quad\; f_3 \\
\begin{array}{c} f_1 \\ f_2 \\ f_3 \end{array}
\begin{bmatrix}
1.00 & 0.14 & 0.33 \\
7.00 & 1.00 & 5.00 \\
3.00 & 0.20 & 1.00
\end{bmatrix}
\end{array}
$$

O_3 = Preserve energy
resources

$$
\begin{array}{c}
\quad f_1 \quad f_2 \quad f_3 \\
\begin{array}{c} f_1 \\ f_2 \\ f_3 \end{array}
\begin{bmatrix}
1.00 & 6.00 & 0.33 \\
0.17 & 1.00 & 0.13 \\
3.00 & 8.00 & 1.00
\end{bmatrix}
\end{array}
\qquad
\begin{array}{l}
f_1 = \text{cost of energy} \\
f_2 = \text{environmental quality} \\
f_3 = \text{energy reserves}
\end{array}
$$

$$
\mathbf{H} = \left[\, \bar{\mathbf{h}}_1 : \bar{\mathbf{h}}_2 : \bar{\mathbf{h}}_3 \,\right] =
\begin{array}{c}
\quad O_1 \qquad O_2 \qquad O_3 \\
\begin{bmatrix}
0.645 & 0.081 & 0.285 \\
0.058 & 0.731 & 0.062 \\
0.297 & 0.188 & 0.653
\end{bmatrix}
\begin{array}{c} f_1 \\ f_2 \\ f_3 \end{array}
\end{array}
$$

Multiplication of this matrix **H** by **B** gives a weighting of the factors based on all of the following:

1. relative influence of actors;
2. relative ranking of objects;
3. relative importance of factors for each objective.

Let us call the result of multiplying **H** by $\bar{\mathbf{B}}$ the vector $\bar{\mathbf{C}}$, which is a weighting of factors:

$$
\bar{\mathbf{C}} = \mathbf{H}\bar{\mathbf{B}} =
\begin{bmatrix}
0.478 \\
0.150 \\
0.372
\end{bmatrix}
\begin{array}{c} f_1 \\ f_2 \\ f_3 \end{array}
$$

Finally, we can extend the analysis to include the importance of a particular factor in realizing a possible future scenario. That is, for each factor what are the pairwise comparisons of the contribution of this factor to the possible future scenarios (S_1 = using oil, S_2 = using coal and S_3 = using nuclear fuel?

F_1 = cost of energy

$$
\begin{array}{c}
\quad S_1 \quad S_2 \quad S_3 \\
\begin{array}{c} S_1 \\ S_2 \\ S_3 \end{array}
\begin{bmatrix}
1.00 & 0.33 & 0.14 \\
3.00 & 1.00 & 0.20 \\
7.00 & 5.00 & 1.00
\end{bmatrix}
\end{array}
$$

F_2 = environmental quality

$$
\begin{array}{c}
\quad S_1 \quad S_2 \quad S_3 \\
\begin{array}{c} S_1 \\ S_2 \\ S_3 \end{array}
\begin{bmatrix}
1.00 & 6.00 & 5.00 \\
0.17 & 1.00 & 0.50 \\
0.20 & 2.00 & 1.00
\end{bmatrix}
\end{array}
$$

F_3 = Level of energy reserves

$$
\begin{array}{cccc}
 & S_1 & S_2 & S_3 \\
S_1 & \begin{bmatrix} 1.00 & 0.11 & 0.33 \\ 9.00 & 1.00 & 6.00 \\ 3.00 & 0.17 & 1.00 \end{bmatrix}
\end{array}
\qquad
\begin{array}{l}
S_1 = \text{use of oil} \\
S_2 = \text{Use of coal} \\
S_3 = \text{use of nuclear}
\end{array}
$$

Let us define the corresponding dominant eigenvectors to the matrices F_1, F_2 and F_3 to be $\overline{I}_1$, $\overline{I}_2$ and $\overline{I}_3$, which when concatenated form the matrix I yielding:

$$
\underline{I} = \overline{I}_1 : \overline{I}_2 : \overline{I}_3 =
\begin{array}{cccc}
 & F_1 & F_2 & F_3 \\
S_1 & \begin{bmatrix} 0.081 & 0.726 & 0.068 \\ 0.188 & 0.102 & 0.770 \\ 0.731 & 0.172 & 0.162 \end{bmatrix} \\
S_2 & \\
S_3 &
\end{array}
$$

Multiplication of I by the vector $\overline{C}$ yields the vector $\overline{D}$ which is the overall weighting of the possible future scenarios (the composite scenario) that is based on the hierarchical weighting of actors' influence, objectives' intensity and factors' importance.

$$
\overline{D} = I\overline{C} =
\begin{bmatrix} 0.173 \\ 0.392 \\ 0.435 \end{bmatrix}
\begin{array}{l} S_1 \\ S_2 \\ S_3 \end{array}
$$

In this particular example, as one might expect, the use of nuclear energy in producing the future's electric power returned the highest weight. We must recall that this result is determined by judgemental priorities throughout the hierarchical structure of the system. The judgements are made in the context of the *decision environment* and therefore, in some sense, measure the response of the actors in the energy environment system to the problems presented in that environment.

In this illustration the present 'state of the world' was taken as the decision environment. This included such factors as recession and a financial dilemma for utilities which emphasizes the cost factor in the judgemental priorities and leads one to suspect that nuclear would come out with the highest weight, particularly in light of the present state of high fuel costs for the fossil alternatives. Indeed, the nuclear alternative did yield the highest weight, but the coal alternative was next (actually fairly close to the nuclear) and, as expected, both coal and nuclear received much higher weights than the oil alternative.

In the next chapter a modification of the hierarchical application of the eigenvalue model will be developed that includes system variables which correspond to the variables used in a model of system process functions in an energy-environment system. The notions of future and composite scenario will be much more rigorously defined in terms of these variables so that ultimately a link can be constructed between a model of system process functions and the hierarchy of policy-making interests in an energy environment system.

3. Policy programming

In the last chapter we examined the structure of energy-environment systems and developed a model for the system process functions in such systems, i.e:

$$X^* = D^*Y$$

We recall that Y in this model is a set of *decision variables* and X^* is a set of *state variables*. We consider a set of decision variables and a set of state variables collectively as a set of *system variables*. Based on the model of system process functions we can now develop some basic definitions that will provide a framework for linking the objectives of policy makers in an energy-environment system with the model of system process functions. The overall linking procedure will be called *policy programming*. In particular, these definitions will be used to first construct a modified hierarchical model of objectives and relative influence of policy makers in an energy-environment system that relates these objectives and influence to a collection of system variables. Several theorems are included to illustrate some important concepts.

3.2. DEFINITIONS

Definition 1. A scenario is a collection of system variables $\mathscr{X} = (Y, X^*)$ where Y is a set of *decision* variables and X^* is a set of *state* variables.

Definition 2 (the principle of consistency). A scenario $\mathscr{X} = (Y, X^*)$ is *con-*

sistent if it satisfies the following:

$$\mathbf{X}^* = \mathbf{D}^*\mathbf{Y},$$

where $\mathbf{D}^*$ is the model of system process functions.

We recall (section 2.4) that $\mathbf{D}^*$ was found by:

$$\mathbf{X} = (\mathbf{I} - \mathbf{A})^{-1}\mathbf{Y}$$
$$\mathbf{X}^* = \mathbf{D}\mathbf{X}, \text{ so } \mathbf{D}^* = \mathbf{D}(\mathbf{I} - \mathbf{A})^{-1}$$

where $(\mathbf{I} - \mathbf{A})^{-1}$ is the Leontief inverse and $\mathbf{D}$ is the matrix of coefficients that relates levels of energy consumption, environmental pollution and employment $(\mathbf{X}^*)$ to total industry outputs $(\mathbf{X})$.

Theorem 1. The scenario $\mathcal{X} = (\mathbf{Y}, \mathbf{X}^*)$ is equivalently defined by $(\mathbf{X}, \mathbf{X}^*)$.

Proof. The range space $\mathbf{X}$ and the domain space $\mathbf{Y}$ are isomorphic.

This implies that the consistency condition can be stated alternately as:

$$\mathcal{X} = \mathbf{T}^*\mathbf{Y} \qquad \mathcal{1} = (\mathbf{X}, \mathbf{X}^*)$$

where $\mathbf{T}^*$ is formed by concatenating $(\mathbf{I} - \mathbf{A})^{-1}$ and $\mathbf{D}^*$:

$$\mathbf{T}^* = \left[\frac{(\mathbf{I} - \mathbf{A})^{-1}}{\mathbf{D}^*}\right] \quad \mathbf{D}^* = \mathbf{D}(\mathbf{I} - \mathbf{A})^{-1}$$

Theorem 2. If $\mathcal{R} = \sum_{i=1}^{n} c_i \mathcal{X}_i$, where $\mathcal{X}_i$, $i - 1, ?, \ldots, n$ are consistent scenarios, then $\mathcal{R}$ is also consistent. c_i, $i = 1, 2, \ldots, n$ are non-negative constants.

Proof. $\mathcal{R}$ is a linear combination of the $\mathcal{X}_i$'s.

$\mathcal{R}$ is of the form $\sum_{i=2}^{n} c_i \left[\frac{(\mathbf{I} - \mathbf{A})^{-1}}{\mathbf{D}^*}\right]\mathbf{Y}_i$

but **A** and **D** are constant so:

$$\mathcal{R} = \left[\frac{(\mathbf{I} - \mathbf{A})^{-1}}{\mathbf{D}^*}\right] \sum_{i=1}^{n} c_i \mathbf{Y}_i = \left[\frac{(\mathbf{I} - \mathbf{A})^{-1}}{\mathbf{D}^*}\right] \sum \mathbf{Y}_i'$$

where $\mathbf{Y}_i' = c_i \mathbf{Y}_i$ or:

$$\mathcal{R} = \left[\frac{(\mathbf{I} - \mathbf{A})^{-1}}{\mathbf{D}^*}\right] \mathbf{Y}_i'' \qquad\qquad \text{where } \mathbf{Y}_i'' = \sum \mathbf{Y}_i'$$

which is the same form as that in *Theorem 2*, hence $\mathcal{R}$ is consistent.

Definition 2. A *future scenario* $\mathcal{X}_k$ is a consistent scenario that describes some future state of the energy-environment system. There are $k = 1, 2, \ldots, K$ such future scenarios.

Definition 4. A *preference scenario* $\mathcal{X}_i$ is a scenario preferred by policy maker i $(i = 1, 2, \ldots, n)$ that is composed by:

$$\mathcal{X}_i = \sum_{k=1}^{K} d_{ik} \mathcal{X}_k$$

where $\mathcal{X}_k$ is a future scenario and d_{ik} is a measure of *relative preference* for that future scenario by policy maker $i(d_{ik} > 0, \sum_{k=1}^{K} d_{ik} = 1)$.

Theorem 3. A *preference scenario* is consistent.

Proof. Direct application of *Theorem 2*.

Definition 5. A cluster $\mathcal{X}_e$ is a group of closely related system variables over which a policy maker $i(i = 1, 2, \ldots, n)$ has a measurable amount of influence c_{ie}, $(\sum_{i=1}^{n} c_{ie} = 1)$. A *complete* set of clusters $(e = 1, 2, \ldots, E)$ constitutes a scenario, $c_{ie} \geq 0$. We note that, in general, $c_{ie} \neq c_{je}$, i.e. policy makers have varying degrees of influence over different clusters of system variables.

Definition 6. A *composite scenario* $\mathcal{R}$ is formed by the complete set of clusters $\mathcal{R}_e$ where:

$$\mathcal{R}_e = \sum_{i=1}^{n} c_{ie} \mathcal{X}_{ie}$$

where $\mathcal{X}_{ie}$ is the cluster (e) of system variables taken from the preferred scenario of policy maker i.

Definition 7. A *reference scenario* $\mathcal{X} = (\mathbf{Y}, \mathbf{X}^*)$ is a consistent scenario that contains the current (in time) state of the energy-environment system being considered.

With the preceding definitions and theorems we now develop a modified hierarchical model based upon the original hierarchical model of chapter 2. This modified model will use the revised definitions of the various *scenarios* in terms of *system variables* just presented. The important notion of *consistency* (*Definition 2*) will be of special concern in determining a composite scenario by this modified hierarchical model.

3.3. THE MODIFIED HIERARCHICAL APPROACH

In chapter 2 the hierarchical eigenvalue model was reviewed. We recall that the purpose of the hierarchical approach is to arrange actors of a policy-making system in a manner so that the priorities and relative influence of these actors can be more easily examined. In this section, we deal more specifically with energy-environment systems and develop a modified hierarchical approach that links the objectives and relative influence of policy makers in such a system to the variables included in the model of process functions outlined in chapter 2.

In this modified hierarchical approach we also attempt to expand the concept of relative influence (*Definition 5*) over system variables from the rather vague notion of overall influence used in chapter 2 to a more realistic one of influence over closely related *clusters* (*Definition 5*) of system variables.

In the modified approach we will make use of the various scenarios defined in the last section. In particular, the judgements in the modified approach will be made in the context of a *reference scenario*. The policy makers judge the relative contribution each of a number of *future scenarios* has toward fulfilling their individual objectives. The result of this process will be a set of *preference scenarios*, one for each policy maker, that are linear combinations of the future scenarios where the dominant future scenario receives more weight than that of the other future scenarios. This preference scenario, for each policy maker, can be viewed as the future which a policy maker would exert his influence in order to attain. Finally, based upon the relative influence that policy makers can bring to bear in order to realize their preference scenarios, a *composite scenario* will result. Let us review specifically how each of these scenarios is derived.

A matrix, called the *future scenario matrix*, $(\mathcal{A})$, can be constructed with entries a_{jk}, each of which is the value of system variable $j(j = 1, 2, \ldots, J)$ corresponding to the kth *future scenario* $(k = 1, 2, \ldots, K)$. This is simply formed by placing K *future scenarios* (*Definition 3*) as columns of a matrix.

The hierarchical structure of policy makers and objectives allows us to examine how the decisions of policy makers shape the future of the energy-environment system through the system process functions. These policy makers make decisions according to their own sets of objectives and the priorities they place on those objectives. We recall that the eigenvalue model utilizes pairwise comparisons of policy maker objectives to yield a normalized vector of priorities placed on the objectives. Let us refer to this procedure as $\mathbf{B}_i \rightarrow \overline{\mathbf{B}}_i$, where $\mathbf{B}_i$ is the matrix of pairwise comparisons of objectives for policy maker i (P_i) and $\overline{\mathbf{B}}_i$ is the corresponding vector of priorities computed by the eigenvalue model. Hence, a vector of priorities on objectives for each policy maker, P_i, $i = 1, 2, \ldots, n$, can be determined.

Policy makers then provide pairwise comparisons of the contribution that the *future scenarios* have toward fulfilling each of their objectives. Let us denote these matrices by $\mathbf{C}_{io}$, an element of which describes the relative desirability of one *future scenario* over another judged by P_i with regard to his objective o. Again, the eigenvalue model can be applied to each $\mathbf{C}_{io}$ to obtain $\overline{\mathbf{C}}_{io}$, the vector of relative desirability of the future scenarios for objective o of P_i. We now form the *matrix of scenario preferences* $\mathcal{C}_i$, the columns of which are the vectors of priorities placed on the future scenarios for all the objectives for policy maker i. Hence, for n policy makers there are n such matrices. The product $\overline{\mathbf{D}}_i = \mathcal{C}_i \cdot \overline{\mathbf{B}}_i$ gives the overall relative desirability of the future scenarios for policy maker i. The *preference scenario* is then given by $\overline{\mathbf{F}}_i = \mathcal{A}\,\overline{\mathbf{D}}_i$, a vector whose values are the system variables, weighted according to the priorities of policy maker i placed on the future scenarios (a *preference scenario* is a linear combination of future scenarios and is, therefore, consistent – see *Theorem 1*).

We can derive the *composite scenario* as follows. First, we divide the system variables into *clusters* (*Definition 5*) $e = 1, 2, \ldots, E$. In a simple case we may wish simply to divide the set of system variables into a few groups, e.g. three: (1) energy related variables, (2) economic variables, and (3) environmental variables. A matrix $\mathbf{G}_e$ is constructed from the pairwise comparisons of influence that each policy maker has over a particular cluster e. Again, the eigenvalue model is employed to determine $\overline{G}_e$, the vector of relative influence of policy makers over e. We recall that $\overline{F}_i$, the preference scenario of P_i, is a collection of system variables; hence it can also be divided into the *clusters* determined above: $\overline{F}_i = [F_{i1}, F_{i2}, \ldots, F_{iE}]$. Finally, the *composite scenario* is found by the collection of the following *clusters* ($\overline{R}_e$ is a *cluster* of the composite scenario $\overline{R}$):

$$\overline{R}_e = [\overline{F}_{1e}, \ldots, \overline{F}_{ne}] \cdot \overline{G}_e, \text{ for } e = 1, 2, \ldots, E \ (a \ complete \ set \ of$$
clusters).

This resulting composite scenario is based upon the relative influence of all policy makers in the energy-environment system, the priority each policy maker places on his individual objectives, and the relative desirability the future scenarios promise toward fulfilling those objectives. Table 3.1 summarizes the modified hierarchical approach and its procedure for finding (1) the composite scenario, (2) the preference scenarios, and (3) the relative influence of policy makers over system variables, all of which will be used later.

There is a problem with this composite scenario. Since the values of policy-maker influence over system variables may be different for each cluster, we cannot be sure that the composite scenario is consistent. This is because it is not necessarily a linear combination of the preference scenarios; hence, we cannot assure consistency by *Theorem 2*.

If $\overline{R}$ is not consistent, it can be viewed as a future to which the policy makers in the energy-environment system would prefer the system move if it were feasible in terms of the system process functions. Hence, this composite scenario constitutes a set of planning targets for system variables that are derived from policy-makers' objectives and relative influence alone. The problem then is to find a consistent scenario that is as close as possible to these planning targets. In the process of determining this *consistent composite scenario* ($\overline{R}^*$) we may be required to alter some of the planning targets. We would like to do this alteration in an efficient manner, preserving the hierarchy of influence developed in the hierarchical model. In other words, the targets established as a result of higher priority objective fulfillment should be preserved while lower priority targets may be relaxed.

Before we develop a method to obtain the consistent composite scenario let us return to the simple example of the hierarchical model presented in chapter 2. However, this time we will use the modified model to determine a *composite scenario* of system variables. Later we will again return to this example to determine the *consistent composite scenario*.

TABLE 3.1. Modified hierarchical model summary.

1. Form $\mathcal{A} = a_{jk}$, where $j = 1, 2, \ldots, J$ (the system variables),
   $$k = 1, 2, \ldots, K \text{ (the future scenarios).}$$
   a_{js} = value of system variable j in the future scenario s. Note that $\mathcal{A}_k$ (*future scenario k*) is a *consistent scenario*.

2. Apply the eigenvalue model to the pairwise comparison matrix of objectives for P_i, $i = 1, 2, \ldots, n$ ($\mathbf{B}_i$) to obtain $\bar{\mathbf{B}}_i$ ($\mathbf{B}_i \rightarrow \bar{\mathbf{B}}_i$); an element of $\bar{\mathbf{B}}_i$ gives the relative dominance of an objective over other objectives of P_i.

3. Apply the eigenvalue model to $\mathbf{C}_{io}$ (the pairwise comparison matrix of *future scenarios* with respect to objective o of P_i) to obtain $\mathbf{C}_{io}$ ($\mathbf{C}_{io} \rightarrow \mathbf{C}_{io}$); an element of $\bar{\mathbf{C}}_{io}$ gives the relative dominance of scenario s for objective o of P_i.

4. Form $\mathbf{D}_i$, an element of which is the relative desirability of a *future scenario* for P_i:
   $$\bar{\mathbf{D}}_i = \mathcal{C}_i \cdot \bar{\mathbf{B}}_i$$

5. Form $\bar{\mathbf{F}}_i$, the *preference scenario* for P_i:
   $$\bar{\mathbf{F}}_i = \mathcal{A} \cdot \bar{\mathbf{D}}_i$$

6. Apply the eigenvalue model to $\mathbf{G}_e$ (the matrix of pairwise comparisons of P_i's influence ($i = 1, 2, \ldots, n$) over system variable *cluster e* (a set of closely related system variables) to obtain $\bar{\mathbf{G}}_e$ ($\mathbf{G}_e \rightarrow \bar{\mathbf{G}}_e$) for $e = 1, 2, \ldots, E$ (all clusters).

7. Divide $\bar{\mathbf{F}}_i$ into *clusters* corresponding to $e = 1, 2, \ldots, E$:
   $$\bar{\mathbf{F}}_i = [\bar{\mathbf{F}}_{i1}, \bar{\mathbf{F}}_{i2}, \ldots, \bar{\mathbf{F}}_{ie}, \ldots, \bar{\mathbf{F}}_{iE}]^\mathsf{T}$$

8. Form the *composite scenario* for each subsystem:
   $$\bar{\mathbf{R}}_e = [\bar{\mathbf{F}}_{1e}, \bar{\mathbf{F}}_{2e}, \ldots, \bar{\mathbf{F}}_{ne}] \cdot \bar{\mathbf{G}}_e$$

   The overall *composite scenario* is simply the collection of $\mathbf{R}_e$'s:
   $$\mathbf{R} = [\mathbf{R}_1, \mathbf{R}_2, \ldots, \mathbf{R}_E]^\mathsf{T}$$

Hierarchical model example (revisited). Consider, as before, three policy makers in a simple energy-environment system:

1. the *government*, which regulates the consumption of energy and the maintenance of environmental quality and influences the level of reserves of energy resources;

2. a private profit-motivated *utility* which generates electricity at the rate demanded by a consuming public and which is subject to the regulatory influence of the government;
3. *consumers* of electrical energy who are concerned about the price of energy as well as the quality of the environment in which they live.

The actors can pursue any or all of the following objectives, some of which will be pursued more vigorously than others:

1. minimizing the cost of producing and consequently consuming energy;
2. maintenance of a high standard of environmental quality;
3. preservation or conservation of precious natural energy resources.

Let us presume that the system variables in the simple energy environment system are the following (in three clusters):

1. energy; $\begin{cases} x_1 = \text{consumption level of oil} \\ x_2 = \text{consumption level of coal} \end{cases}$

2. economic; $x_3 = \text{cost of electricity produced}$

3. environmental $x_4 = \text{level of SO}_2 \text{ emissions}$

Suppose now that the above energy-environment system is presented with three alternatives for producing electric power in the future. These three alternatives constitute the possible *future scenarios* of the energy environment system:

1. use of oil-fired plants;
2. use of coal-fired plants;
3. use of nuclear plants.

First we construct a *future scenario matrix* (A) that contains the values of system variables for each of the future scenarios. Let us consider the A matrix for our simple energy environment system to be as shown in table 3.2.

In order to construct the *composite scenario* based on the preferences of the actors in the system, we first apply the eigenvalue model to the objectives pursued by each actor. Each actor pursues the defined objectives with varying degrees of emphasis. Pairwise comparison of the defined objectives of each actor can accordingly be made to determine the priority with which an actor views each of the sets of objectives.

TABLE 3.2. Future scenarios: example

Clusters	Variables	Scenarios		
		Oil	Coal	Nuclear
Energy	Oil consumption	4×10^{15}Btu	3×10^{15}Btu	2×10^{15}Btu
	Coal consumption	2×10^{15}Btu	4×10^{15}Btu	1×10^{15}Btu
Economy	Cost of electrical energy	5¢/kwh	4¢/kwh	3¢/kwh
Environment	SO_2 emissions	8×10^6 tons/year	9×10^6 tons/year	6×10^6 tons/year

In this simple hypothetical case for instance, the government may in some decision environment emphasize the minimizing of energy cost to the maintenance of environmental quality and preservation of natural resources by, say, a factor of three. Accordingly, the emphasis of maintenance of environmental quality to minimization of energy cost would be one-third (in the terms of the eigenvalue model).

Let us call the matrix of comparisons of objectives for the government actor B_1, that for the utility B_2, and that for consumers B_3 in this simple illustration. A typical set of matrix constructions of these comparisons might be:

$$B_1 = \text{government}$$
$$\begin{array}{c} \\ O_1 \\ O_2 \\ O_3 \end{array} \begin{array}{ccc} O_1 & O_2 & O_3 \\ \begin{bmatrix} 1.00 & 3.00 & 3.00 \\ 0.33 & 1.00 & 0.33 \\ 0.33 & 3.00 & 1.00 \end{bmatrix} \end{array}$$

$$B_2 = \text{utility}$$
$$\begin{array}{c} \\ O_1 \\ O_2 \\ O_3 \end{array} \begin{array}{ccc} O_1 & O_2 & O_3 \\ \begin{bmatrix} 1.00 & 8.00 & 4.00 \\ 0.13 & 1.00 & 0.17 \\ 0.25 & 6.00 & 1.00 \end{bmatrix} \end{array}$$

$$B_3 = \text{consumer}$$
$$\begin{array}{c} \\ O_1 \\ O_2 \\ \\ O_3 \end{array} \begin{array}{ccc} O_1 & O_2 & O_3 \\ \begin{bmatrix} 1.00 & 2.00 & 3.00 \\ 0.50 & 1.00 & 2.00 \\ 0.33 & 0.50 & 1.00 \end{bmatrix} \end{array}$$

$O_1 =$ minimizing energy cost
$O_2 =$ maintaining environmental quality
$O_3 =$ preserving natural resources

The dominant eigenvectors of B_1, B_2 and B_3, which we can call $\bar{B}_1$, $\bar{B}_2$ and $\bar{B}_3$, respectively, describe the relative priorities of the objectives for each actor:

$$\bar{\mathbf{B}}_1 = \begin{bmatrix} 0.584 \\ 0.135 \\ 0.281 \end{bmatrix} \begin{matrix} O_1 \\ O_2 \\ O_3 \end{matrix} \qquad \bar{\mathbf{B}}_2 = \begin{bmatrix} 0.691 \\ 0.060 \\ 0.249 \end{bmatrix} \begin{matrix} O_1 \\ O_2 \\ O_3 \end{matrix} \qquad \bar{\mathbf{B}}_3 = \begin{bmatrix} 0.540 \\ 0.297 \\ 0.163 \end{bmatrix} \begin{matrix} O_1 \\ O_2 \\ O_3 \end{matrix}$$

Similarly, we must construct the pairwise comparison matrices of the desirability of the future scenarios with respect to each objective for each actor. In our simple system we have three objectives that are common to all the actors. Hence, we must find the relative weights of each of the three future scenarios with regard to each of the three objectives for each actor, a total of nine pairwise comparison matrices. For the purpose of illustration, let us presume that the pairwise comparison matrices of the scenarios for each of the electric utility actor's objectives are the following:

$\mathbf{C}_{21}$ = lower cost of energy

	S_1	S_2	S_{23}
S_1	1.00	0.33	0.14
S_2	3.00	1.00	0.20
S_3	7.00	5.00	1.00

$\mathbf{C}_{22}$ = maintain environmental quality

	S_1	S_2	S_3
S_1	1.00	6.00	5.00
S_2	0.17	1.00	0.50
S_3	0.20	2.00	1.00

$\mathbf{C}_{23}$ = preserve level of energy reserves

	S_1	S_2	S_3
S_1	1.00	0.11	0.33
S_2	9.00	1.00	6.00
S_3	3.00	0.17	1.00

S_1 = use of oil
S_2 = use of coal
S_3 = use of nuclear

Let us define the corresponding dominant eigenvectors of the matrices $\mathbf{C}_{21}$, $\mathbf{C}_{22}$, and $\mathbf{C}_{23}$ to be $\bar{\mathbf{C}}_{21}$, $\bar{\mathbf{C}}_{22}$, and $\mathbf{C}_{23}$, which when concatenated forms the matrix $\mathbf{C}_2$:

$$\mathbf{C}_2 = [\bar{\mathbf{C}}_{21} : \bar{\mathbf{C}}_{22} : \bar{\mathbf{C}}_{23}] = \begin{matrix} S_1 \\ S_2 \\ S_3 \end{matrix} \begin{matrix} O_1 & O_2 & O_3 \\ \begin{bmatrix} 0.081 & 0.726 & 0.068 \\ 0.188 & 0.102 & 0.770 \\ 0.731 & 0.172 & 0.162 \end{bmatrix} \end{matrix}$$

This matrix C_2 contains the relative weights of the three future scenarios for each objective of the electric utility. Multiplication of C_2 by the vector $\bar{B}_2$ gives the overall relative desirability of the scenarios for the electric utility based on the relative priorities of its objectives:

$$\bar{D}_2 = C_2\bar{B}_2 = \begin{array}{c} S_1 \\ S_2 \\ S_3 \end{array} \begin{array}{ccc} O_1 & O_2 & O_3 \\ \left[\begin{array}{ccc} 0.081 & 0.726 & 0.068 \\ 0.188 & 0.102 & 0.770 \\ 0.731 & 0.172 & 0.162 \end{array}\right] \end{array} \left[\begin{array}{c} 0.691 \\ 0.060 \\ 0.249 \end{array}\right] = \left[\begin{array}{c} 0.116 \\ 0.328 \\ 0.556 \end{array}\right] \begin{array}{c} S_1 \\ S_2 \\ S_3 \end{array}$$

Note that for the electric utility, the nuclear scenario is the most preferred, which is expected.

We can calculate this D_i vector (weighting of scenarios) for each of the actors in the simple energy-environment system:

Government actor

C_{11} = lower cost of energy

$$\begin{array}{c} S_1 \\ S_2 \\ S_3 \end{array} \begin{array}{ccc} S_1 & S_2 & S_3 \\ \left[\begin{array}{ccc} 1.0 & 0.25 & 0.20 \\ 4.0 & 1.0 & 0.33 \\ 5.0 & 3.0 & 1.0 \end{array}\right] \end{array}$$

C_{12} = maintain environmental quality

$$\begin{array}{c} S_1 \\ S_2 \\ S_3 \end{array} \begin{array}{ccc} S_1 & S_2 & S_3 \\ \left[\begin{array}{ccc} 1.0 & 5.0 & 2.0 \\ 0.20 & 1.0 & 0.33 \\ 0.50 & 3.0 & 1.0 \end{array}\right] \end{array}$$

C_{13} = preserve energy resources

$$\begin{array}{c} S_1 \\ S_2 \\ S_3 \end{array} \begin{array}{ccc} S_1 & S_2 & S_3 \\ \left[\begin{array}{ccc} 1.0 & 0.2 & 0.5 \\ 5.0 & 1.0 & 3.0 \\ 2.0 & 0.33 & 1.0 \end{array}\right] \end{array}$$

$$C_1 = \begin{array}{c} S_1 \\ S_2 \\ S_3 \end{array} \begin{array}{ccc} O_1 & O_2 & O_3 \\ \left[\begin{array}{ccc} 0.108 & 0.582 & 0.122 \\ 0.339 & 0.109 & 0.648 \\ 0.553 & 0.309 & 0.230 \end{array}\right] \end{array}, \quad \bar{D}_1 = C_1\bar{B}_1 = \left[\begin{array}{c} 0.176 \\ 0.395 \\ 0.429 \end{array}\right] \begin{array}{c} S_1 \\ S_2 \\ S_3 \end{array}$$

Consumer

$$
\begin{array}{ccc}
\mathbf{C}_{31} & \mathbf{C}_{32} & \mathbf{C}_{33} \\
\begin{array}{c} \\ S_1 \\ S_2 \\ S_3 \end{array}
\begin{array}{ccc}
S_1 & S_2 & S_3 \\
\begin{bmatrix} 1.0 & 0.33 & 0.25 \\ 3.0 & 1.0 & 0.5 \\ 4.0 & 2.0 & 1.0 \end{bmatrix}
\end{array}
&
\begin{array}{ccc}
S_1 & S_2 & S_3 \\
\begin{bmatrix} 1.0 & 4.0 & 3.0 \\ 0.25 & 1.0 & 0.5 \\ 0.33 & 2.0 & 1.0 \end{bmatrix}
\end{array}
&
\begin{array}{ccc}
S_1 & S_2 & S_3 \\
\begin{bmatrix} 1.0 & 0.25 & 0.33 \\ 4.0 & 1.0 & 4.0 \\ 3.0 & 0.25 & 1.0 \end{bmatrix}
\end{array}
\end{array}
$$

$$
\mathbf{C}_3 = \begin{array}{c} S_1 \\ S_2 \\ S_3 \end{array}
\begin{array}{ccc}
O_1 & O_2 & O_3 \\
\begin{bmatrix} 0.122 & 0.625 & 0.113 \\ 0.320 & 0.136 & 0.652 \\ 0.558 & 0.238 & 0.235 \end{bmatrix}
\end{array},
\quad
\mathbf{\bar{D}}_3 = \mathbf{C}_3 \mathbf{\bar{B}}_3 =
\begin{bmatrix} 0.270 \\ 0.319 \\ 0.410 \end{bmatrix}
\begin{array}{c} S_1 \\ S_2 \\ S_3 \end{array}
$$

Finally, we have determined the relative desirability of each scenario for each actor ($\mathbf{\bar{D}}_1$, $\mathbf{\bar{D}}_2$, $\mathbf{\bar{D}}_3$) based on the relative priorities of each one's objectives. We can concatenate these vectors to form the matrix $\mathbf{D}$:

$$
\mathbf{D} = [\mathbf{\bar{D}}_1 : \mathbf{\bar{D}}_2 : \mathbf{\bar{D}}_3] =
\begin{array}{c} S_1 \\ S_2 \\ S_3 \end{array}
\begin{array}{ccc}
A_1 & A_2 & A_3 \\
\begin{bmatrix} 0.176 & 0.116 & 0.270 \\ 0.395 & 0.328 & 0.319 \\ 0.429 & 0.556 & 0.410 \end{bmatrix}
\end{array}
$$

Multiplication of $\mathcal{A}$, which we recall is the matrix of the reference values of the state variables for each of the future scenarios, by $\mathbf{D}$, results in a matrix $\mathbf{F}$, the columns of which give the *preferred composite scenarios* (*preference*) of each of the actors:

$$
\mathbf{F} = \mathcal{A}\mathbf{D} =
\begin{array}{c} f_1 \\ f_2 \\ f_3 \\ f_4 \end{array}
\begin{array}{ccc}
S_1 & S_2 & S_3 \\
\begin{bmatrix} 4 & 3 & 2 \\ 2 & 4 & 1 \\ 5 & 4 & 3 \\ 8 & 9 & 6 \end{bmatrix}
\end{array}
\cdot
\begin{array}{c} S_1 \\ S_2 \\ S_3 \end{array}
\begin{array}{ccc}
A_1 & A_2 & A_3 \\
\begin{bmatrix} 0.176 & 0.116 & 0.270 \\ 0.395 & 0.328 & 0.319 \\ 0.429 & 0.556 & 0.410 \end{bmatrix}
\end{array} =
$$

$$
\begin{array}{c} f_1 \\ f_2 \\ f_3 \\ f_4 \end{array}
\begin{array}{ccc}
A_1 & A_2 & A_3 \\
\begin{bmatrix} 2.75 & 2.56 & 2.86 \\ 2.36 & 2.1 & 2.33 \\ 3.75 & 3.56 & 3.86 \\ 7.54 & 7.22 & 7.49 \end{bmatrix}
\end{array}
$$

In order to determine the *composite scenario*, we must account for the relative influence that each of the actors has over the three clusters of indicators (energy, economic and environmental) in the simple energy environment system. We can construct a pairwise comparison matrix of actors' influence over each of the three clusters:

$$
\begin{array}{cc}
\text{energy } G_1 & \text{economy } G_2 \\[2pt]
\begin{array}{c}
 \\ A_1 \\ A_2 \\ A_3
\end{array}
\begin{array}{ccc}
A_1 & A_2 & A_3 \\
\begin{bmatrix} 1.0 & 5.0 & 3.0 \\ 0.2 & 1.0 & 4.0 \\ 0.33 & 0.25 & 1.0 \end{bmatrix}
\end{array}
&
\begin{array}{c}
 \\ A_1 \\ A_2 \\ A_3
\end{array}
\begin{array}{ccc}
A_1 & A_2 & A_3 \\
\begin{bmatrix} 1.0 & 7.0 & 5.0 \\ 0.143 & 1.0 & 0.167 \\ 0.2 & 6.0 & 1.0 \end{bmatrix}
\end{array}
\end{array}
$$

$$
\begin{array}{c}
\text{environment } G_3 \\[2pt]
\begin{array}{c}
 \\ A_1 \\ A_2 \\ A_2
\end{array}
\begin{array}{ccc}
A_1 & A_2 & A_3 \\
\begin{bmatrix} 1.0 & 2.0 & 5.0 \\ 0.5 & 1.0 & 3.0 \\ 0.2 & 0.33 & 1.0 \end{bmatrix}
\end{array}
\end{array}
\qquad
\begin{array}{l}
A_1 = \text{government} \\
A_2 = \text{electric utility} \\
A_3 = \text{consumer}
\end{array}
$$

The dominant eigenvectors of these matrices ($\overline{G}_1$, $\overline{G}_2$ and $\overline{G}_3$, respectively) give the weighting of influence each actor has over energy, economic and environmental factors in the simple energy environment system:

$$
G = [\overline{G}_1 : \overline{G}_2 : \overline{G}_3] =
\begin{array}{c}
 \\ A_1 \\ A_2 \\ A_3
\end{array}
\begin{array}{ccc}
\text{Energy} & \text{Economy} & \text{Environment} \\
\begin{bmatrix} 0.640 & 0.708 & 0.582 \\ 0.242 & 0.062 & 0.309 \\ 0.114 & 0.230 & 0.109 \end{bmatrix}
\end{array}
$$

Finally, we can take the group of energy factors from F (a partition of F) which we recall describes the preferred values of energy factors for each of the actors and multiply by $\overline{G}_1$, which is the vector describing the relative influence each actor has over energy factors. The result is a vector ($\overline{R}_1$) of the values of energy factors based on the preferred values of each actor as well as the influence each actor has in determining those factors. In this simple case we can determine the resulting factors for all of the factor groups (energy, economic and environmental) as follows:

	A_1	A_2	A_3		e_1	e_2	e_3
energy $\{f_1$	2.75	2.56	2.86	A_1	0.640	0.640	0.708	0.582
$\{f_2$	2.36	2.1	2.23	A_2	0.242	0.242	0.062	0.309
economic $\{f_3$	3.75	3.56	3.86	A_3	0.114	0.114	0.230	0.109
environmental $\{f_4$	7.54	7.22	7.49					

The diagonal elements of the resulting matrix form the vector $\overline{R}$, the composite scenario (off diagonal elements are meaningless):

$$\mathbf{R} = \begin{bmatrix} 2.7 \\ 2.27 \\ 3.76 \\ 7.43 \end{bmatrix} \begin{array}{l} \times \ 10^{15} \ \text{Btu oil consumption} \\ \times \ 10^{15} \ \text{Btu coal consumption} \\ \text{¢/kwh cost of electricity} \\ \times \ 10^{6} \ \text{tons/year SO}_2 \ \text{emissions} \end{array}$$

Note that in this simple example the preferred scenario is closer to the nuclear scenario than the other two future scenarios. This is perhaps due to the fact that the decision environment, in the context of which all the judgemental matrices are formed, was chosen to be severe economic recession so that the need for cheap energy becomes emphasized. The emphasis on cheap energy then overshadows other objectives, maintenance of environmental quality and preservation of energy reserves. Since the preferred scenarios all emphasize the nuclear scenario, the resulting composite scenario likewise tends toward the nuclear future scenario.

Even in this example, had we chosen a suitable model of system process functions, we could not guarantee that the composite scenario would be consistent with these process functions. We, therefore, review a method, goal programming, that will lead us to a *consistent composite scenario*.

3.4. GOAL PROGRAMMING

Goal programming is a multiobjective mathematical programming procedure that turns out to be well suited to the problem of determining a *consistent composite scenario* which appeared in the last section. In this section we will review the goal programming method to see why it appears to be such a promising approach. Let us take a general multiobjective optimization problem, i.e.:

$$\min \ \mathcal{F} \ (\mathbf{x}) = \{f_1(\mathbf{x}), f_2(\mathbf{x}), \ldots, f_n(\mathbf{x})\} \ \text{subject to} \ \mathbf{x} \in \ T$$

where $f_i(\mathbf{x})$ is one of n objectives and T is the feasible region of solutions $\mathbf{x}$.

Goal programming requires that the decision maker(s) set goals or targets for each of the n objectives. The 'optimal' solution is then defined as the one that minimizes the deviations from established goals. If there are several objectives, priorities can be placed on the objectives.

Suppose we choose a vector of values as goals for the objectives in the above problem: $\hat{\mathscr{F}} = \{\hat{f}_1, \hat{f}_2, \ldots, \hat{f}_n\}$. The goal programming formulation of the general multiobjective problem then becomes:

$$\min \| \mathscr{F}(\mathbf{x}) - \hat{\mathscr{F}} \| \text{ subject to } \mathbf{x} \in T$$

The symbol $\|.\|$ denotes any norm, e.g. the sum of the absolute values of the deviations of individual objective functions from their prescribed goals. Other norms may be more realistic if there is diminishing value to the full attainment of goals (diminishing marginal utility). A normal procedure for formulating goal programming problems is to introduce deviational variables:

$$\min \|\mathbf{d}\| \text{ where } \mathbf{d} = \{d_1, d_2, \ldots, d_n\}$$

subject to

$$\mathscr{F}(\mathbf{x}) - \mathbf{d} = \hat{\mathscr{F}}$$

$$\mathbf{x} \varepsilon T$$

In application measurement of goal attainment in goal programming usually takes the form of the ℓp metric.[1] In policy programming we shall use the ℓ_1 metric since the model of system process functions developed for energy-environment systems in chapter 2 is a linear model. In addition, solutions to ℓ_1 formulated goal programming problems turn out to be easily found by modified simplex procedure similar to that of linear programming.

This modified simplex procedure, which will be discussed shortly, will handle multiple prioritized objectives at two levels:
1. establishment of preemptive priority classes where the simplex procedure attempts to satisfy high priority objectives before moving on to lower priority ones – this allows incommensurate goals to be included in the same problem.
2. within a single preemptive priority class, multiple objectives can be weighted (as in parametric linear programming).

Both of these types of objective priorities will be important in the policy programming application since the number of priority classes in (1) will be determined by the important clusters of system variables identified

---

1. Excellent discussions of norms used in goal programming are given in Lane [210] and Charnes and Cooper [175].

in the modified hierarchical model of the last section. Within each cluster the relative dominance of a policy maker's influence over that cluster will provide the weights for the objective functions within a given priority class.

The linear ℓ_1 formulation of goal programming was originally developed by Charnes and Cooper [175] and has been applied in many areas such as manpower planning (Charnes et al. [178]), media planning (Charnes et al. [180]), academic planning (Lee [211]), aggregate production planning (Jaaskelainen [203]), environment resource planning (Panagiotakopoulos [217]) and a host of others. In practice, the only applications of goal programming have been with the linear formulation, but some conceptual work on nonlinear goal programming has been done (Ijiri [201]).[2]

Let us now review the procedure for solving ℓ_1 formulated linear goal programs by means of a modified simplex algorithm.

Solution of linear goal programming problems. The linear formulation of goal programming can be viewed as an extension of linear programming. As one might expect, a simplex solution procedure has been developed for goal programming that is quite similar to that of linear programming, with several distinct differences.[3] The first difference, of course, is the problem formulation. Let us consider the following example, a profit maximization problem, to illustrate the formation of linear goal programs. The linear programming (LP) formulation of the example is given by:

$$\max Z = c_1 x_1 + c_2 x_2$$
$$\text{subject to } x_1 + x_2 \leq b_1$$
$$x_1 \qquad \leq b_2$$
$$x_2 \leq b_3$$
$$x_1, x_2 \quad \geq 0 \qquad\qquad (3.4.1)$$

There are actually two objectives in this linear program: The first is implicit, i.e. the solution $x^* = (x_1^*, x_2^*)$ must lie within the feasible region defined by the constraint equations. The second objective is the explicit objective which is to maximize the objective function.

The example can be translated into a linear goal programming problem (LGP) by introducing deviational variables (d_1, d_2 and d_3) that denote deviations from the right-hand side values of the constraint equations and another (d_4) that denotes the deviation from an established goal (g) for the

2. Lee [211] summarizes Ijiri's work.
3. The best reference for the modified simplex for goal programming is Lee [211].

objective function. To account for the priorities placed on objectives, we assign a preemptive priority factor (p_1) to the feasibility objective (implicit) and another factor (p_2) to the explicit maximization objective.

In goal programming, we think of the first objective as minimizing the positive deviations of the values of the constraint equations from their right-hand side values and the second objective as minimizing negative deviation of the value of the objective from its established target g. Hence, the LGP formulation of the example can be given by:

$$\min Z = p_1(d_1^+ + d_2^+ + d_3^+) + p_2 d_4^-$$

subject to:
$$x_1 + x_2 + d_1^- - d_1^+ = b_1$$
$$x_1 + d_2^- - d_2^+ = b_2$$
$$x_2 + d_3^- - d_3^+ = b_3$$
$$c_1 x_1 + c_2 x_2 + d_4^- - d_4^+ = g \qquad (3.4.2)$$

Note that for the solution of the LGP problem to be identical to that of the LP problem $p_1 > p_2$ in order to assure feasibility before optimality. The values p_1 and p_2 are ordinal values and the simplex algorithm, as discussed below, treats them in order of priority. Objectives within a given priority level can be weighted, as in parametric linear programming.

An important point to consider in goal programming is that in order to solve equation (3.4.1) (the LP problem), a feasible region defined by the constraints is required. If the feasible regions of (3.4.1) and (3.4.2) were both described by A in fig. 3.1, then the problem is solvable by both LP and LGP. However, if this feasible region were described by B then the problem would be unsolvable by linear programming but goal programming would yield a 'best compromise' solution in terms of the established goals and priorities placed on objectives. In addition, the LGP problem illustrated by (3.4.2) is trivial since it deals with only one explicit goal. Larger problems become much more complicated as the numbers of variables and objectives increase.

We recall that the simplex method for linear programming (Dantzig [185]) involves the following five basic steps.

1. formulate the simplex tableau by introducing necessary artificial and slack variables which form the initial tableau.
2. determine the new variable that will enter the basis, i.e. find the variable that will contribute the most to the objective function.

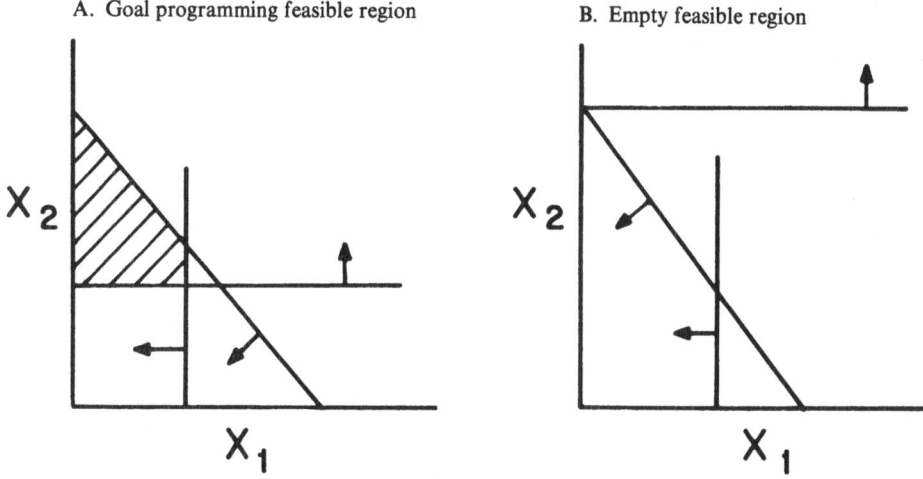

FIG. 3.1. Feasible regions.

3. determine the variable leaving the basis, i.e. find the variable that constrains the solution the most.
4. determine the new basic feasible solution.
5. determine whether the new solution is or is not optimal; if it is not, repeat steps two to five until the optimum is found.

The basic steps for the modified simplex procedure of goal programming are quite similar to that of linear programming except that the criteria for determining entering and exiting variables are designed to include the notion of goal attainment rather than optimality in the LP sense. The basic steps to the modified simplex procedure are the following

1. Formulate the simplex tableau. Original problem variables are considered 'choice' variables while a set of 'deviational' variables are introduced to determine goal attainment. Negative deviational variables form the initial basis and are assumed at the origin. Objectives are listed in order of priority; i.e. a number of rows are included in the tableau, one for each priority level. The goals or targets of these objectives (which, in addition to the ordinal priorities, is additional information required in goal programming) are used to measure the goal attainment.
2. Determine the new entering variable. In GP this is identified by the column of the tableau whose coefficient in the highest unattained

priority level row decreases the unattainment of the highest goal at the fastest rate. This is accomplished by first identifying the highest unattained priority level, then determining the column which has the highest positive $Z_j - C_j$ value.

3. Determine the existing variable by finding the element (row) of column found in step two, whose value when divided by C_j is a minimum.

4. Determine the new basic feasible solution which is the result of a row reduction of the entire tableau pivoting on the element defined in steps two and three.

5. If all Z_j values of each priority level are zero the solution is optimal. If they are not, then determine if there are negative $Z_j - C_j$ values at a higher priority level in the same column of unattained objectives, i.e. the positive $Z_j - C_j$ elements of the unattained objective. If this is the case, then the solution is 'preferred' or 'best compromise.' If the solution is neither optimal or preferred, the steps two to five are repeated until an optimum or preferred solution is reached.

Example: the goal programming solution. Consider the following simple goal programming problem:

$$\min Z = p_1 d_1^- + p_2 d_4^+ + p_3(5d_2^- + 3d_3^-) + p_4 d_1^+$$

$$
\begin{aligned}
x_1 + x_2 + d_1^- \quad\quad\quad\quad\quad -d_1^+ \quad\quad\quad &= 80 \\
x_1 \quad\quad\quad +d_2^- \quad\quad\quad -d_2^+ \quad\quad\quad &= 70 \\
x_2 \quad\quad\quad +d_3^- \quad\quad\quad -d_3^+ \quad\quad &= 45 \\
x_1 + x_2 \quad\quad\quad\quad +d_4^- \quad\quad\quad\quad -d_4^+ &= 90
\end{aligned}
$$

$$(3.4.3)$$

We note that there are four preemptive priority levels. The third level has two sub-objectives with relative weights 5 and 3, respectively. Graphically, this problem is shown in figure 3.2. If we attempt to fulfill each of the objectives in order of priority, we can observe the solution graphically. Any point northeast of the line segment AB in the figure satisfies the first priority goal (minimize d_1^-); any point then in the cross-hatched region satisfies the first two priority goals (minimize d_1^- and d_4^+); upon adding the third priority goal (the higher weighted one in the third priority level – minimize d_2^-) only the line segment CD satisfies all three objectives; finally adding the fourth goal (minimize d_3^-) moves the solution to the point D without compromising higher priority objectives. The addition of

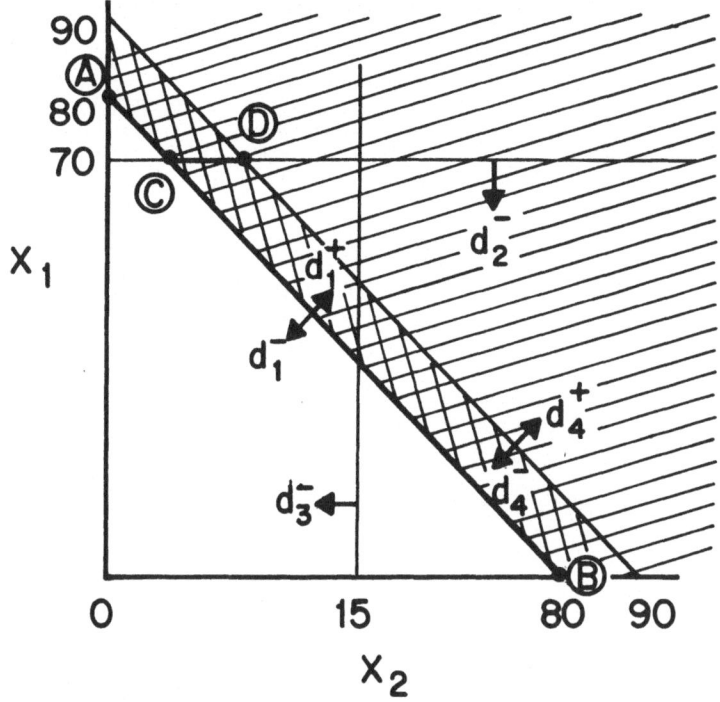

FIG. 3.2. Goal programming example.

the last goal (minimize d_1^+) does not change the solution point without compromising higher priority goals.

The simplex method of goal programming can be easily applied to this example. The initial tableau for the example is shown in table 3.3. The subsequent iterations are shown in table 3.4. Note that in the final tableau for positive values in the objective function rows (d_4^- in the p_4 row and d_4^+ in the p_3 row), there are negative values below in higher priority objective rows.

Goal programming and the policy programming framework. Despite several problems in application, which will appear in chapter 6, goal programming has a number of features that make it well suited to the policy programming framework. These features are the following:

1. Goal programming is computationally efficient. The linear formulation can be solved by the simplex algorithm just described which is quite similar to the simplex method of linear programming.

TABLE 3.3. Goal programming example: simplex method (initial tableau).

C_j					p_1	$5p_3$	$3p_3$		p_4	p_2
	V	C	x_1	x_2	d_1^-	d_2^-	d_3^-	d_4^-	d_1^+	d_4^+
p_1	d_1	80	1	1	1	0	0	0	-1	0
$5p_3$	d_2^-	70	1	0	0	1	0	0	0	0
$3p_3$	d_3^-	45	0	1	0	0	1	0	0	0
	d_4^-	90	1	1	0	0	0	1	0	-1
	p_4	0	0	0	0	0	0	0	-1	0
	p_3	485	5	3	0	0	0	0	0	0
	p_2	0	0	0	0	0	0	0	0	-1
	p_1	80	1	1	0	0	0	0	-1	0

Note: initial target levels are found by:
$$p_1 = 1 \times 80$$
$$p_2 = 0$$
$$p_3 = (5 \times 70) + (3 \times 45) = 485$$
$$p_4 = 0$$

2. Goal programming was designed to come 'as close as possible to a set of simultaneously unattainable goals.' The solution algorithm attempts to satisfy high priority goals first and then attempts to achieve the lower priority targets without compromising the already established higher priority targets.
3. Finally, the goal programming model can easily utilize the model of system process functions for an energy-environment system developed in chapter 2. This we will see in what follows.

3.5. DETERMINING THE CONSISTENT COMPOSITE SCENARIO

In the last section we found how goal programming could be used to obtain a solution that is 'as close as possible' to a set of planning targets while satisfying a given constraint set. Let us now see how goal programming can be used for finding the *consistent composite scenario* we sought in section 3.3.

First, we recall that the three principal results of the modified hierarchi-

TABLE 3.4. Goal programming example: simplex method (subsequent tableaux)

C_j					p_1	$5p_3$	$3p_3$		p_4	p_2
	V	C_j	x_1	x_2	d_1^-	d_2^-	d_3^-	d_4^-	d_1^-	d_4^-
p_1	d_1^-	10	0	1	1	−1	0	0	−1	0
	x_1	70	1	0	0	1	0	0	0	0
$3p_3$	d_3^-	45	0	1	0	0	1	0	0	0
	d_4^-	20	0	1	0	−1	0	1	0	−1
	p_4	0	0	0	0	0	0	0	−1	0
	p_3	135	0	3	0	−5	0	0	0	0
	p_2	0	0	0	0	0	0	0	0	−1
	p_1	10	0	1	0	−1	0	0	−1	0

C_j					p_1	$5p_3$	$3p_3$		p_4	p_2
	V	C_i	x_1	x_2	d_1^-	d_2^-	d_3^-	d_4^-	d_1^+	d_4^+
	x_2	10	0	1	1	1	0	0	−1	0
	x_1	70	1	0	0	1	0	0	0	0
$3p_3$	d_3^-	35	0	0	−1	1	1	0	1	0
	d_4^-	10	0	0	−1	0	0	1	1	−1
	p_4	0	0	0	0	0	0	0	−1	0
	p_3	105	0	0	−3	−2	0	0	3	0
	p_2	0	0	0	0	0	0	0	0	−1
	p_1	0	0	0	−1	0	0	0	0	0

C_j					p_1	$5p_3$	$3p_3$		p_4	p_2
	V	C_i	x_1^-	x_2^-	d_1^-	d_2^-	d_3^-	d_4^-	d_1^+	d_4^+
	x_2	20	0	1	0	−1	0	1	0	−1
	x_1	70	1	0	0	1	0	0	0	0
$3p_3$	d_3^-	25	0	0	0	1	1	−1	0	1
p_1	d_1^-	10	0	0	−1	0	0	1	1	−1
	p_4	10	0	0	−1	0	0	1	0	−1
	p_3	75	0	0	0	−2	0	−3	0	3
	p_2	0	0	0	0	0	0	0	0	−1
	p_1	0	0	0	−1	0	0	0	0	0

cal model of section 3.3 were the following:

1. a set of preference scenarios $\overline{F}_i$, $i = 1, 2, \ldots, n$ (one for each policy maker).
2. a set of vectors giving the *relative influence* of policy makers over system variable *clusters*: $\overline{G}_e$, $e = 1, 2, \ldots, E$ (a *complete* set of clusters E).
3. a *composite scenario* $\overline{R}$.

The *composite scenario* (*Definition 6*) contains a set of *state variables* X^* and a set of *decision variables* Y:

$$\overline{R} = (Y, X^*)$$

We recall that the condition of consistency was given by: $X^* = D^*Y$, where D^* is a model of the system process functions. We recall also that D^* is found by the following:

$$X = (I - A)^{-1}Y$$
$$X^* = DX$$

where D is the total intensity matrix and A is the matrix of technical co-efficients (section 2.4). So:

$$D^* = D(I - A)^{-1}$$

These equations can also be written:

$$Y = (I - A)X$$
$$X^* = DX$$

Let us define the matrix B to be the concatenation of $(I - A)$ and D:

$$B = \left[\frac{(I - A)}{D} \right]$$

Hence:

$$\overline{R} = B X$$

We recall that in order to make $\overline{\mathbf{R}}$ consistent, we must alter one or more of the planning targets, values of system variables given in the composite scenario. In order to allow deviation both above and below the planning targets, we can introduce deviational variables:

$$\mathbf{B\,X} + \mathbf{d}^- - \mathbf{d}^+ = \overline{\mathbf{R}}$$

where $\mathbf{d}^-$ and $\mathbf{d}^+$ are vectors of negative (slack) and positive (surplus) deviational variables.

The system of equations now looks much like that of the constraint set in a goal programming problem. Collective objectives of policy makers (those that would be fulfilled by the composite scenario if it were consistent) that are relevant to the system process functions can be expressed in terms of the deviational variables. Individual objectives of policy makers could also be expressed in terms of deviational variables:

$$\mathbf{B\,X} + \mathbf{d}^- - \mathbf{d}^+ = \overline{\mathbf{F}}_i \qquad\qquad i = 1, 2, \ldots, n.$$

The complete model including individual and collective objectives would then be given by:

$$\min \mathbf{P}_+ \mathbf{d}^+ + \mathbf{P}_- \mathbf{d}^-$$
$$\text{subject to} \quad \mathbf{B\,X} + \mathbf{d}_1^- - \mathbf{d}_1^+ = \overline{\mathbf{R}}$$
$$\mathbf{B\,X} + \mathbf{d}_{i+1}^- - \mathbf{d}_{i+1}^+ = \overline{\mathbf{F}}_i \qquad\qquad i = 1, 2, \ldots, n$$

The preemptive priority levels are established for each cluster of system variables. The weights of objectives within a particular priority level are determined by the relative influence policy makers have over the cluster indicated by that level, i.e.:

$$\mathbf{P}_e\{\overline{\mathbf{G}}_{e1}(\mathbf{d}_{e2}^- \mid \mathbf{d}_{e2}^+) + \overline{\mathbf{G}}_{e2}(\mathbf{d}_{e3}^- + \mathbf{d}_{e3}^+) + \cdots$$
$$+ \overline{\mathbf{G}}_{en}(\mathbf{d}_{e(n+1)}^- + \mathbf{d}_{e(n+1)}^+)\}$$
$$\text{where } \overline{\mathbf{G}}_e = [\overline{\mathbf{G}}_{e1}, \overline{\mathbf{G}}_{e2}, \ldots, \overline{\mathbf{G}}_{en}].$$

$d_{e(i+1)}$ is the set of deviational variables (related to system variables in cluster e) that describe deviation from the individual targets (preference scenario) of policy maker i. The index of the deviational variables is $i + 1$ rather than i because $\mathbf{d}_{e1}$ is the set of deviational variables for the collective objectives (*composite scenario*).

The goal programming problem results in a consistent scenario by *Definition 2* since $\mathbf{X}'$, the solution to the goal programming problem, produces consistent values for $\mathbf{Y}$ and $\mathbf{X}^*$.

$$\mathbf{X}^* = \mathbf{D}\mathbf{X}'$$
$$\mathbf{Y} = (\mathbf{I} - \mathbf{A})\,\mathbf{X}'$$

Practical problems involved in the solution of GP problems in the policy programming context will be discussed in chapter 6. The principal problem to reckon with is the enormous size of the simplex tableaus that are required in solving policy programming problems.

We now return to the example of the modified hierarchical approach (the example of section 3.3) and see how goal programming can be used to develop a *consistent composite scenario*.

The hierarchical model example (revisited again). We recall that the composite scenario developed by the modified hierarchical application of the eigenvalue model was the following (example of section 3.3):

$$\overline{\mathbf{R}} = \begin{bmatrix} 2.7 \\ 2.27 \\ 3.76 \\ 7.43 \end{bmatrix} \begin{array}{l} \times\ 10^{15}\ \text{Btu oil consumed} \\ \times\ 10^{15}\ \text{Btu consumed coal} \\ \text{¢/kwh price of electricity} \\ \text{million tons of SO}_2\ \text{emissions} \end{array}$$

For purposes of the example, let us append two more variables on to the composite scenario: the final demands presented to two industries supplying goods and services to the region of interest (arbitrary amounts – say $5 million for industry A and $5 million for industry B). As a result the $\overline{\mathbf{R}}$ vector now looks as follows:

$$\overline{\mathbf{R}} = \begin{bmatrix} 5 \\ 5 \\ 2.7 \\ 2.27 \\ 3.76 \\ 7.43 \end{bmatrix} \begin{array}{l} \text{million \$ (industry A)} \\ \text{million \$ (industry B)} \\ 10^{15}\ \text{Btu oil consumed} \\ 10^{15}\ \text{Btu coal consumed} \\ \text{¢/kwh price of electricity} \\ \text{million tons of SO}_2\ \text{emissions} \end{array}$$

We presume also for purposes of the example, that the regional economy can be described by the following two-sector input-output model:

	A	B	Final demand	Total output
Industry A	3	2	5	10
Industry B	1	7	2	10
Value added	6	1		
Total input	10	10		

The technical coefficients matrix is easily found ($a_{ij} = Z_{ij}/X_j$):

$$A = \begin{bmatrix} 0.3 & 0.2 \\ 0.1 & 0.7 \end{bmatrix} \quad (I - A) = \begin{bmatrix} 0.7 & -0.2 \\ -0.1 & 0.3 \end{bmatrix}$$

where Z_{ij} is an element of the transactions matrix and X_j is an element of the total outputs vector. Also, let us presume that the levels of energy consumption, cost of electricity and SO_2 emissions in the region are related to industrial outputs by the following:

$$X^* = DX$$

where X = the vector of total outputs of industry A and industry B, respectively. So:

$$D = \begin{bmatrix} 0.2 & 0.3 \\ 0.1 & 0.4 \\ 0.5 & 0.2 \\ 0.5 & 1.1 \end{bmatrix} \quad \begin{array}{l} 10^{15} \text{ Btu oil/\$ output} \\ 10^{15} \text{ Btu coal/\$ output} \\ (\text{¢/kwh})/\$ \text{ output} \\ \text{tons } SO_2/\$ \text{ output} \end{array}$$

The input-output model and the total impact coefficients (D) form a representation of the system process functions for the region. Given this information it becomes very clear that $\overline{R}$, the composite scenario, is *inconsistent*. The equilibrium point described by the input-output model is given by:

$$X = (I - A)^{-1}Y = \begin{bmatrix} 13.15 \\ 21.05 \end{bmatrix}$$

where $Y = (5, 5)^T$ – the final demands of R.

The corresponding values of the other system variables $\mathbf{X}^*$ (energy consumption, electricity price and SO_2 emissions) are computed by:

$$\mathbf{X}^* = \mathbf{DX} = \begin{bmatrix} 9.02 \\ 9.77 \\ 10.74 \\ 29.68 \end{bmatrix}$$

The values of the variables in $\mathbf{X}^*$ are not the same as the corresponding values in $\overline{\mathbf{R}}$. How then can we revise these planning targets (given in $\overline{\mathbf{R}}$) so that they represent a consistent scenario, i.e. so both of the following hold?

$$\mathbf{X} = (\mathbf{I} - \mathbf{A})^{-1}\mathbf{Y}$$
$$\mathbf{X}^* = \mathbf{DX}$$

or as presented in chapter 2:

$$\mathcal{X} = (\mathbf{X}, \mathbf{X}^*) = \mathbf{T}^*\mathbf{Y}$$

$$\text{where } \mathbf{T}^* = \begin{bmatrix} (\mathbf{I} - \mathbf{A})^{-1} \\ \hline \mathbf{D}^* \end{bmatrix} \quad \mathbf{D}^* = \mathbf{D}(\mathbf{I} - \mathbf{A})^{-1}$$

In revising the planning targets, we wish to preserve the priority hierarchy established in setting the planning targets in the first place. In other words, we seek to come up with a consistent scenario that comes as close as possible to the planning targets given in the composite scenario ($\overline{\mathbf{R}}$).

The answer, of course, as we found earlier is to invoke the goal programming procedure with $\overline{\mathbf{R}}$ forming the target levels in the goal program and the relationships given by $\mathbf{A}$ and $\mathbf{D}$ providing the constraint equations. The goal program for this example might be formed as follows:

$$\min Z = \mathbf{P}_- \mathbf{d}^- + \mathbf{P}_+ \mathbf{d}^+$$
subject to $\mathbf{BX} + \mathbf{ID}^+ + \mathbf{ID}^- = \mathbf{b}$

where in this example $\mathbf{B} = \begin{bmatrix} (\mathbf{I} - \mathbf{A}) \\ \hline \mathbf{D} \end{bmatrix}$ and $\mathbf{b} = \overline{\mathbf{R}}$

If we assign the achievement of the established final demand levels the highest priority level (P_1), the energy consumption factors the next level (P_2), the electricity price the next (P_3), and finally achieving the target

TABLE 3.5. Policy programming: example (case 1).

Initial tableau

C_i					P_1	P_1	P_2	P_2	P_3	P_4	$3P_1$	$3P_1$	P_2	P_2	P_3	P_4
	V	C	x_1	x_2	d_1^-	d_2^-	d_3^-	d_4^-	d_5^-	d_6^-	d_1^+	d_2^+	d_3^+	d_4^+	d_5^+	d_6^+
P_1	d_1^-	5.00	0.70	0.20	1	0	0	0	0	0	-1	0	0	0	0	0
P_1	d_2^-	5.00	-0.10	0.30	0	1	0	0	0	0	0	-1	0	0	0	0
P_2	d_3^-	2.70	0.20	0.30	0	0	1	0	0	0	0	0	-1	0	0	0
P_2	d_4^-	2.27	0.10	0.40	0	0	0	1	0	0	0	0	0	-1	0	0
P_3	d_5^-	3.76	0.50	0.20	0	0	0	0	1	0	0	0	0	0	-1	0
P_4	d_6^-	7.43	0.50	1.10	0	0	0	0	0	1	0	0	0	0	0	-1
	P_4	7.43	0.50	1.10	0	0	0	0	0	0	0	0	0	0	0	-2
$Z_j - C_j$	P_3	3.76	0.50	0.20	0	0	0	0	0	0	0	0	0	0	-2	0
	P_2	4.97	0.30	0.70	0	0	0	0	0	0	0	0	-2	-2	0	0
	P_1	10.00	0.60	0.10	0	0	0	0	0	0	-4	-4	0	0	0	0

Final tableau

	x_1	13.16	1	0	1.58	1.05	0	0	0	0	-1.58	-1.05	0	0	0	0
P_2	d_4^+	7.47	0	0	0.37	1.58	0	-1	0	0	-0.37	-1.58	0	1	0	0
P_4	d_6^+	22.31	0	0	1.37	4.58	0	0	0	-1	-1.37	-4.58	0	0	0	1
P_2	d_3^+	6.25	0	0	0.47	1.32	-1	0	0	0	-0.47	-1.32	1	0	0	0
	x_2	21.05	0	1	0.53	3.68	0	0	0	0	-0.53	-3.68	0	0	0	0
P_5	d_5^+	7.03	0	0	0.89	1.26	0	0	-1	0	-0.89	-1.26	0	0	1	0
	P_4	22.31	0	0	1.37	4.58	0	0	0	2	-1.37	-4.58	0	0	0	0
$Z_j - C_j$	P_3	7.03	0	0	0.89	1.26	0	0	-2	0	-0.89	-1.26	0	0	0	0
	P_2	13.71	0	0	0.84	2.89	-2	-2	0	0	-0.84	-2.89	0	0	0	0
	P_1	0.00	0	0	-1.00	-1.00	0	0	0	0	-3.00	-3.00	0	0	0	0

$$\bar{R}^* = \left[\frac{(I-A)}{D}\right] X' = \begin{bmatrix} 5.0 \\ 5.0 \\ 8.95 \\ 9.74 \\ 10.79 \\ 29.74 \end{bmatrix} \qquad X' = (13.15, 21.05)^T$$

SO_2 emissions level the last priority level (P_4), the goal programming objective function might be given the following:

$$\min Z = P_1(d_1^- + d_2^- - 3d_1^+ + 3d_2^+) + P_2(d_3^- + d_4^- + d_3^- + d_4^+) + P_3(d_5^- + d_5^+) + P_4(d_6^3 + d_6^+)$$

The corresponding simplex tableau is shown in table 3.5. Upon solution by the modified simplex algorithm (the final tableau is also shown in table 3.5) we see that the solution is equivalent to the $\underline{X}$ computed from $X = (I - A)^{-1}Y$ and $X^* = DX$ since $\mathcal{X} = (X, X^*)$. Why? The highest priority action (P_1) was to insure macroeconomic equilibrium $(I - A)^{-1}Y = X$. If we force this requirement, then the only possible consistent scenario is given by $\mathcal{X}$, since $(I - A)^{-1}Y$ is only one point. However, suppose we alter the priority structure. For instance, the relative priority of objectives and influence might dictate that we place the first priority on minimizing SO_2 emissions over the established target. The goal programming problem might then be:

$$\min Z = P_1 d_6^+ + P_2(3d_3^+ + 2d_4^+) + P_3(d_1^- + d_1^+ + d_2^- + d_2^+) + P_4 d_5^+$$
subject to $BX + Id^- - Id^+ = b$

The corresponding initial and final tableaus are shown in table 3.6. Note that in this case the solution values of the system variables are:

$$\overline{R}^* = BX' = \begin{bmatrix} 5.00 \\ 3.10 \\ 2.50 \\ 2.05 \\ 4.64 \\ 7.43 \end{bmatrix} \begin{array}{l} 10^6 \text{ \$ industry A} \\ 10^6 \text{ \$ industry B} \\ 10^{15} \text{ Btu oil} \\ 10^{15} \text{ Btu coal} \\ \text{¢/kwh} \\ 10^6 \text{ tons } SO_2 \end{array}$$

The highest priority objective of meeting the minimum·emissions target was met only at the considerable expense of industry B's output. This indicates a very important conclusion. If the target levels of final demand represented the fraction of materials that would be purchased by, for instance, an energy park development project in the region were those given in R (\$5 million for each industry), then with present technology in the region (described by A and D) the pollution target levels cannot be met. The bill of goods presented to regional industry must be reduced to that of the second case in order to meet the total emissions target.

We can note from the final tableau (table 3.6) that all targets were met except those in the third priority level which were the macroeconomic equilibrium conditions given by the input-output model. The solution points of both examples are shown graphically in figure 3.3.

TABLE 3.6. Policy programming: example (case 2).

Initial tableau

C_i				p_3	p_3					p_3	p_3	$3p_2$	$2p_2$	p_4	p_1
V	C	x_1	x_2	d_1^-	d_2^-	d_3^-	d_4^-	d_5^-	d_6^-	d_1^+	d_2^+	d_3^+	d_4^+	d_5^+	d_6^+
$p_3\ d_1^-$	5.00	0.70	0.20	1	0	0	0	0	0	−1	0	0	0	0	0
$p_3\ d_2^-$	5.00	−0.10	0.30	0	1	0	0	0	0	0	−1	0	0	0	0
d_3^-	2.70	0.20	0.30	0	0	1	0	0	0	0	0	−1	0	0	0
d_4^-	2.27	0.10	0.40	0	0	0	1	0	0	0	0	0	−1	0	0
d_5^-	3.76	0.50	0.20	0	0	0	0	1	0	0	0	0	0	−1	0
d_6^-	7.43	0.50	1.10	0	0	0	0	0	1	0	0	0	0	0	−1
p_4	0.00	0.00	0.00	0	0	0	0	0	0	0	0	0	0	−1	0
p_3	10.00	0.60	0.10	0	0	0	0	0	0	−2	−2	0	0	0	0
p_2	0.00	0.00	0.00	0	0	0	0	0	0	0	0	−3	−2	0	0
p_1	0.00	0.00	0.00	0	0	0	0	0	0	0	0	0	0	0	−1

Final tableau

V	C	x_1	x_2	d_1^-	d_2^-	d_3^-	d_4^-	d_5^-	d_6^-	d_1^+	d_2^+	d_3^+	d_4^+	d_5^+	d_6^+
x_1	8.03	1	0	1.26	0	0	0	0	0.23	−1.26	0	0	0	0	−0.23
$p_3\ d_2^-$	4.87	0	0	0.30	1	0	0	0	−0.22	0.30	−1	0	0	0	0.22
$p_3\ d_3^-$	0.16	0	0	−0.08	0	1	0	0	−0.29	0.08	0	−1	0	0	0.29
d_4^-	0.23	0	0	0.10	0	0	1	0	−0.34	−0.10	0	0	−1	0	0.34
x_2	3.10	0	1	−0.57	0	0	0	0	0.80	0.57	0	0	0	0	−0.80
$p_4\ d_5^+$	0.88	0	0	0.52	0	0	0	−1	0.28	−0.52	0	0	0	1	−0.28
p_4	0.88	0	0	0.52	0	0	0	−1	0.28	−0.52	0	0	0	0	−0.28
p_3	4.87	0	0	−0.70	0	0	0	0	−0.22	−1.30	−2	0	0	0	0.22
p_2	0.00	0	0	0.00	0	0	0	0	0.00	0.00	0	−3	−2	0	0.00
p_1	0.00	0	0	0.00	0	0	0	0	0.00	0.00	0	0	0	0	−1.00

$$\bar{\mathbf{R}}^* = \left[\frac{(\mathbf{I} - \mathbf{A})}{\mathbf{D}}\right]\mathbf{X}' = \begin{bmatrix} 5.0 \\ 3.10 \\ 2.50 \\ 2.05 \\ 4.64 \\ 7.43 \end{bmatrix} \qquad \mathbf{X}' = (8.03, 3.11)^T$$

3.6. METHODOLOGY SUMMARY

To recapitulate briefly, the result of section 2.3 was a formal description of policy-making systems, extended from the basic concepts of general systems theory to include a hierarchy of objective seeking policy makers (purposeful subsystems). These policy makers influence an overall

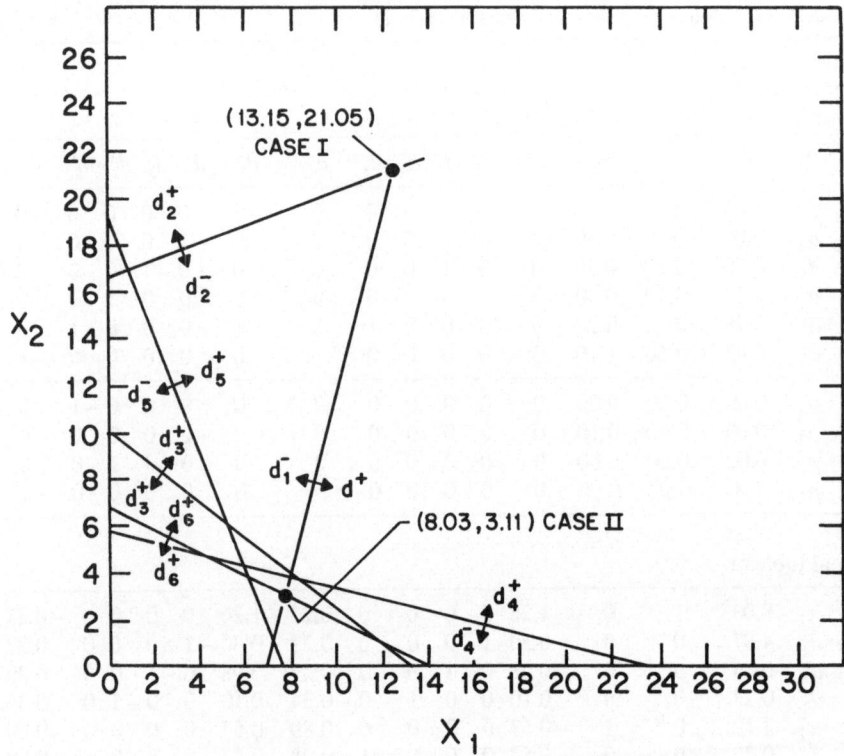

FIG 3.3. Policy programming example: solution points.

system which is controlled by a set of system process functions. In order to operationalize this formalism, hierarchical application of the eigenvalue prioritization model was employed. The model was modified to account for policy makers' influence over individual *clusters* of system variables. The principal results of the modified model were: (1) a set of preference scenarios each of which describes a future state to which a policy maker would prefer the system evolve; (2) a measure of relative influence (a normalized weighting vector) that policy makers have over clusters of system variables; and (3) a *composite scenario* that describes a future state to which the system might evolve if system process functions and constraints were not present, i.e. a state that depends upon the relative influence of policy makers in achieving preferred outcomes alone.

In order to account for these process functions, some methodology was borrowed from the theory of mathematical programming. In particular, the existence of a target state leads quite naturally to goal programming, a method that seeks to minimize underachievement of predetermined goals or targets. The value of an objective function is compared to a target value to assess the degree of goal attainment. In the case of multiple objectives, higher priority objectives, those of policy makers with greater influence, are fulfilled first and then attempts are made to fulfill lower priority objectives without compromising those already achieved. Of course, this mathematical programming procedure is carried out subject to some constraint set, which in this case is a representation of the process functions and constraints of the system of interest. The result of solving the mathematical programming problem then is a scenario, described by the resultant vector of system variables, that accounts not only for the relative influence of the system's policy makers, but also for the effect of relevant system process functions and constraints on the system performance. In fact, the degree of this effect can be observed, in some sense, by comparing the composite scenario with the consistent composite scenario. Similarly, the consistent composite scenario can also be compared with the various preferred scenarios to measure again, in some sense, the degree of effect a particular policy maker has on the system performance. The policy programming framework is summarized in figure 3.4.

An important limitation. One important limitation that has not been discussed or treated by the policy programming framework is that of the problems involved in policy implementation. Indeed, it is assumed that a formulated policy (posed in the model as a number of objective functions) is both feasible and implementable. Such may not be the case. The process of viewing the problem as has been done so far might be called the 'forward' process. Alternatively, viewing the problem first in terms of the problems a policy maker might encounter in implementing his policy might be termed the 'backward' process. Using this backward mode, one can refine his judgements for the forward mode and so forth until consistency is reached in both directions. This iterative process insures a much better measure of consistency in the judgements made in terms of the eigenvalue model. This refinement was not included here. Alexander and Saaty [234] are currently doing research in this very important area with an eye toward incorporating these concepts into the hierarchical prioritization model.

We now turn to Part Two of this study which applies the policy programming framework to the specific problem of energy park development. The results of this application will contribute both to a better understanding of both the policy programming approach and of the problems that might be encountered in energy park development in a region with multiple and

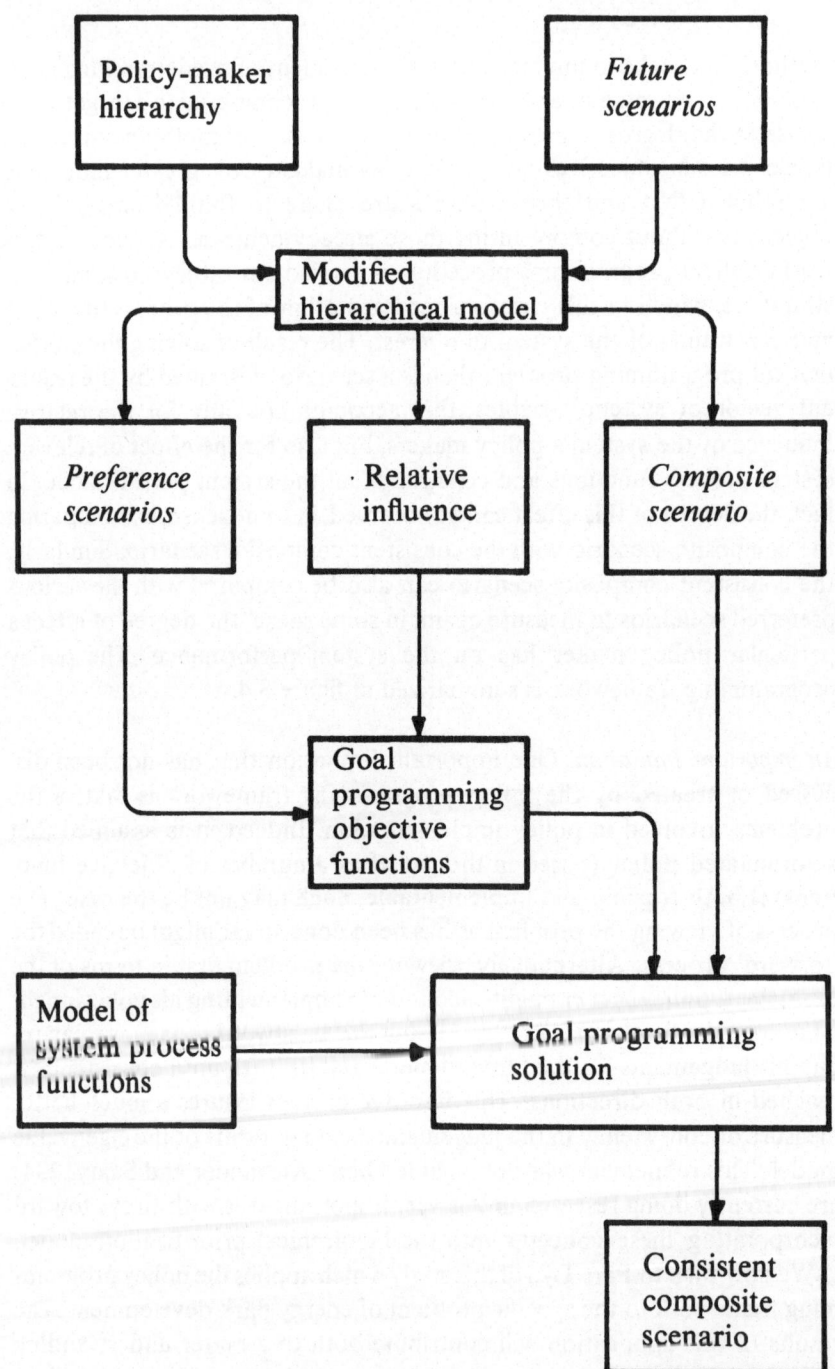

FIG 3.4. Policy programming: summary.

conflicting policy interests. Note that policy programming is certainly not limited to the problem considered in Part Two. However, the energy park problem is rich enough in terms of broad, large-scale planning features so that it serves to illustrate the approach quite well.

Part Two

Planning for energy park development

4. The energy park concept

4.1. INTRODUCTION

In recent years the concept of clustering large amounts of electrical generating capacity (10–50 thousand megawatts) into one geographic site has been suggested as an alternative to dispersed siting of electric power plants. Such large concentrations of power generation have been called energy parks or energy centers. Interest in this energy park concept has been generated by increasing concern over environmental protection and land use. These concerns have made dispersed siting of power plants much more difficult and time consuming to undertake. As a result, utilities have been prompted to consider seriously the energy park concept as a method for meeting future electric power demands, particularly when they are forced to consider sites that are long distances from primary load centers.

In this chapter the basic energy park concept as it has evolved in recent years will be reviewed. In addition, many of the facts that have surfaced to both support and reject the concept as a viable future alternative for the electric power industry will be considered. The problem of planning an energy park in a specific region has all of the characteristics of the long-range planning problems that policy programming was designed to handle. In the next chapter, the development of an energy park at a specific site will be examined using the policy programming framework constructed in Part One.

Two principal avenues of approach have evolved in the consideration of the energy park concept:

1. *Nuclear energy centers.* An energy park composed entirely of nuclear generating capacity has been a focus of study of a number of federal agencies. In particular, the Energy Reorganization Act of 1974 (42 USC 5847, sections 206 and 207) requires that the Nuclear Regulatory Commis-

sion (NRC) complete an evaluation of nuclear energy centers (NEC) in cooperation with federal, state and local authorities. This effort includes NEC's both with and without integrated fuel cycle facilities, the major goal of the former being to centralize fission product (nuclear fuels, wastes, etc) handling facilities. With the projected significant increases in installed nuclear generating capacity in this country over the next several decades (see fig. 4.1), the NRC wishes to investigate whether or not nuclear energy centers can improve safety and reliability in handling nuclear material.

An NEC with an integrated fuel cycle facility (IFCF) would more than likely reduce movement and storage of radioactive materials outside the site boundaries. Alternatively, an NEC without an IFCF would still allow organization of larger fuel shipments than those from dispersed facilities, possibly leading to fewer shipments and, therefore, promoting better security, higher efficiency and increased safety. However, associated with development of NEC's is a host of problems, technological, environmental, economic and social in nature, all of which must be resolved before the energy park can become a reality. The NRC has divided its efforts into the investigation of first *feasibility*, i.e. the study of technological feasibility with regard to technical and environmental aspects involved in siting, construction and operation of energy centers. The second consideration is that of *practicality* or the review of socio-economic impact, jurisdictional questions, financing, national security and safeguards (see NUREG-0001 [110], volume 3). A principal area of concern for the NRC is what the government's role or degree of involvement in planning for energy park development should be if energy parks are found to be feasible and prac-

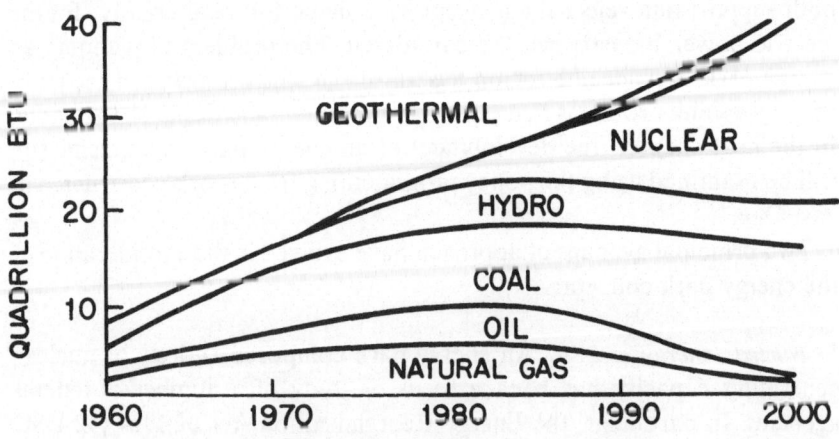

FIG. 4.1. Projected energy use by U.S. electric utilities.

tical. Many of these considerations appear to be highly site-dependent, a fact which has resulted in a study which is part of NRC's response to the requirements of section 207 of the Energy Reorganization Act (NECSS [108]).

2. *Hybrid energy parks.* The other avenue of approach to energy parks has been the consideration of hybrid energy parks that include both fossil and nuclear powered generation. This approach has been the focus of a number of private utilities and federal agencies.[1] However, much less effort in the government investigations is being devoted to this hybrid energy park concept than to nuclear energy centers. The most active interest in this hybrid energy park concept, outside the Federal government, has been in Pennsylvania where the Pennsylvania Energy Park Development Group (made up of members from three Pennsylvania utilities) has investigated the potential of energy parks at a number of sites (see [85]).

In these utility studies, as in the NRC investigations, much attention has been focused on assessing technical feasibility of energy parks. A comprehensive site selection study was recently completed by Gilbert Associates [69] outlining candidate energy park sites in the commonwealth. The Pennsylvania Energy Park Development Group's present conception of a hybrid park is 10,000 megawatts, 50 percent of which would be fossil-fired generation and 50 percent of which would be nuclear fired. The hybrid concept is viewed as more practical by utilities, particularly in Pennsylvania, where political pressures obligate the use of coal, which is one of that state's most plentiful resources (see Saeger [92]).

The shift away from an all-nuclear energy center generally precludes the addition of IFCF in the park design if the installed capacity is below 20,000 megawatts of nuclear generation. As on-site IFCF becomes practical only with a very large nuclear generation complement.

As in the NEC concept, the issues of practicality (socioeconomic, jurisdictional, etc) with regard to hybrid parks are just now beginning to receive attention. The Pennsylvania State University [87] has recently completed a study highlighting some of the key issues concerning practicality of energy parks in the commonwealth.

Table 4.1 shows the various types of energy park designs presently being considered in the U.S. They range from small hybrid energy parks with a fossil and nuclear complement of 5,000 megawatts each (perhaps feasible

---

1. For example FEA and NSF. See General Electric Company [68] and Pennsylvania Energy Park Development Group [85].

TABLE 4.1. The energy park concept.

Nuclear energy centers	Generation installed
Small power center	10–20,000 MW nuclear
Large power center	20–48,000 MW nuclear
Large power center	
with integrated fuel cycle facility (see below)	

Hybrid energy parks	
Small park	5000 MW fossil/5000 MW nuclear
Large park	5000–10,000 MW fossil/10,000–30,000 MW nuclear
Integrated park	variety of fossil, nuclear, hydro (pumped storage), and alternative generation

Integrated fuel handling facilities	
Nuclear fuel reprocessing	
Nuclear fuel fabrication	
Low and high level nuclear waste management	

Source: Edison Electric Institute

by 1985) to a monstrous 48,000 megawatt nuclear park with its own IFCF, which would require an investment of perhaps more than forty billion dollars over several decades (certainly not feasible until the 1990's).

The issues involved in the assessment of technical feasibility of various energy park designs will be discussed in some detail shortly. In general, smaller energy park designs (around 10,000 megawatts) appear technically feasible but the degree of feasibility is highly site dependent, sensitive to both the size and composition of the park. Perhaps even more site dependent are many of the issues of practicality. These issues will be identified discussed shortly.

Evolution of the energy park concept. The original idea of energy parks as a reactor siting concept should be attributed to Alvin Weinberg [116], former director of Oak Ridge National Laboratory, who first proposed the concept in 1971. Prior to this time, nuclear energy centers had been considered for their potential in stimulating regional economic development (see NUREG-0001 [110], Meier [74], ORNL-4290 [83], and USAEC [107]). Current studies of the concept are being undertaken in order to assess whether or not energy parks more effectively deal with future load growth demands, as mentioned earlier, as well as to determine the benefits of centralizing nuclear material handling facilities.

The early studies, i.e. prior to 1970, were primarily concerned only with nuclear energy centers. These included a cooperative effort between the International Atomic Energy Agency, the USAEC and the Mexican government that investigated the feasibility of constructing nuclear power and water desalting plants in the southwestern United States. This study concluded as early as 1968 that $10GW_e$ NEC's were technically feasible in this region with growth to $50GW_e$ by 1995 (see NUREG-0001 [110]). Also in 1968, Oak Ridge National Laboratory studies [83] indicated that the potential technical and economic benefits of NEC's were sufficient to warrant further investigation. Several specific site studies for Puerto Rico and the Middle East in 1970 supported these conclusions (see USAEC [107], ORNL-4481 [82]).

Weinberg's original argument for nuclear parks was fourfold:[2]

1. *To promote reactor safety* – he argued that by clustering reactor facilities one would be able to mobilize resources more easily to cope with reactor accidents which would be impractical at smaller facilities.
2. *To increase control over inventory and shipment of nuclear materials* – he saw NEC's as an efficient mechanism for centralizing fission product handling facilities.
3. *To reduce the number of dispersed decommissioned reactor sites.*
4. *To promote waste heat utilization* – nuclear power plants give enormous amounts of waste heat that might become more useful in the quantities that would be made available at NEC's.

Due to Weinberg, the nuclear energy center concept was seriously considered by FEA and AEC in 1972 and 1973 (see USAEC [102]). Dixy Lee Ray, chairman of the AEC in 1973, initiated the original AEC work on NEC's as part of Project Independence (USAEC [104], [102]). John Sawhill, FEA Administrator at the time, also viewed NEC's as a means of responding to public concerns about nuclear safety. In addition, a number of electric utilities began to review the concept as an answer to future load growth demands. Notably, Pacific Power and Light investigated a site on Lake Roosevelt for up to ten nuclear units; Florida Power and Light evaluated two sites of approximately $12GW_e$ proposed generation each (see Sniley et al. [96]); and the Pennsylvania Energy Park Development Group commissioned Gilbert Associates to produce an energy park site identification study in Pennsylvania (see Gilbert Associates [69]). FEA has also commissioned

---

2. This appears as a sequence of articles: Weinberg [116], [118], [119], Weinberg and Burwell [120], Weinberg [117] and NUREG-0001 [110], volume 2.

four site-specific studies at Camp Gruber, Oklahoma; Glasgow Air Force Base, Montana; Tuscarosa Basin, New Mexico; and Central Michigan.

All of these studies commenced well before the recently concluded Nuclear Energy Site Survey mentioned earlier which is the latest and most comprehensive study of the energy park concept. A good summary of the legislative action leading up to the NECSS is given in the NRC final report on nuclear energy center siting [111]. As a result of NRC and individual electric utility interest, the energy park concept has been vigorously discussed in recent years but results from the site-specific studies have only just begun to appear. Moreover, most of these site-specific studies have generally been restricted to technical aspects of feasibility and environmental impact.

The electric power industry and the energy park. The energy park concept actually follows quite logically, if not justifiably, from historical trends in the electric power industry in the United States. The industry itself has grown remarkably since its beginning in 1879 when Thomas Edison and the English inventor George Lane-Fox first proposed transmitting electrical energy to customers. The industry really did not begin to grow until George Westinghouse and William Stanley introduced the first AC constant potential generation and the transformer (based on the British patents of Gaulard and Gibbs) which demonstrated the practicality and advantages of the present AC distribution system. As a result of the efforts of Westinghouse, Stanley and others, by the early 1900's several long transmission lines were in service and the present public utility concept was fairly well established.[3]

As early as 1925, the idea of consolidation of electric generation, transmission, and distribution facilities was a heated issue known commonly at that time as 'giant power development' (see Cooke [63]). The world's first intercompany power pool, the Pennsylvania-New Jersey Interconnection began operations in 1928, which marked the first large-scale integrated electric power operations project demonstrating the greater reliability and economy offered by power system interconnection. However, even then (circa 1925), social and political questions inhibited attempts to consolidate power facilities. Some of the same questions that presently face the energy park concept, which is itself another more extensive form of consolidation, were posed in 1925: such as who should

---

3. A history of the early electric power industry is given in Novick [79].

have jurisdiction over 'giant power' facilities, what kind of cooperative arrangements between utilities are acceptable (the anti-trust question), and whether or not society would tolerate gigantic overhead transmission lines. The difference is, of course, that 'giant power' in the 1920's referred to a few megawatts; today one thinks of a few gigawatts.

Since the 1920's the power industry has grown to be the largest U.S. industry in terms of capital investment, accounting for some 15.7 percent of total business expenditure of new equipment in the nation.[4] A trend toward consolidation is readily apparent as the number of electric utility companies has actually been dropping since the early years of the industry. Extensive interconnection, as well as vertical integration, have contributed the favorable levels of reliability and economic efficiency the industry has enjoyed until recent years. In addition, the trend has been to move to larger and larger individual generating units to realize economies of scale. However, this trend has been met in recent years by considerable criticism and resistance from environmental concerns, particularly in the areas of siting and operating new generation facilities.

Let us now review the technical *feasibility* issues involved in the energy park concept which suggests as ultimate a form of consolidating electric generating facilities by current standards as giant power did in the 1920's. Next, issues of *practicality* related to the concept will be reviewed. Many current studies have concluded that the energy park is technically feasible at many sites. However, social and political consideration, particularly public acceptance, have precluded many sites from consideration. In Pennsylvania, the original procedure for siting energy parks was to choose the most socially and politically practical site from a set of technically acceptable ones. Due to overwhelming public resistance to many of the proposed sites, the order of siting criteria appears to have been reversed, i.e., choose the best technical site from a set of socially and politically acceptable ones (Mulford [76]).

4.2. TECHNICAL FEASIBILITY

We next review the technical considerations involved in conventional dispersed power plant siting and how these considerations relate to the proposed alternative of siting energy parks or nuclear energy centers. Many of these considerations would not differ qualitatively from con-

4. A summary of capital investment figures is given in NUREG-0001 [110].

ventional siting requirements (only in magnitude). However the problem becomes more complex when hybrid parks are considered since the requirements for nuclear power stations are quite different from those of fossil-fired stations, but we shall see that this is not necessarily objectionable and indeed may be beneficial.

We can classify the technical considerations involved in power plant siting into the following categories: (1) *physical site requirements,* including location and geology, environmental impact, transportation access and land use; (2) *plant design*, including generation, transmission and safety-reliability. In addition, some new considerations for large-scale facilities such as those designed to minimize common-mode failure should be included.

Let us examine each briefly, relate each to the energy park concept and, finally, discuss the overall technical feasibility of energy parks.

1. *Physical site requirements.* The physical requirements of a nuclear power plant and a fossil fueled power plant are quite different. Accordingly, in addition to the much greater magnitude of various site requirements in an energy park design, more complexity is introduced when both the physical requirements of nuclear generation are included.

Proximity to load centers. Historically, a prime requirement of power plants has been that they be located as close as possible to primary load centers in order to achieve maximum economy in transmitting the generated power to those load centers. However, of late, requirements with respect to environmental considerations, such as air quality control and aesthetics, have placed greater restraints on development of fossil plant sites near load centers. As late as 1968 siting experts predicted that 'as experience is gained in the design, construction and operation of nuclear plants the use of locations nearer population centers is possible.'⁵ This optimism has not been justified as yet and, as a result, recent concern over nuclear safety has forced utilities to consider remote sites for nuclear as well as fossil plants. At long distances from load centers transmission considerations, discussed below, suggest centralizing generation facilities, i.e. an energy park.

Hydrology. Perhaps the most restricting physical screening factor in power plant siting is that of securing a dependable cooling source. Of the order of 25 cubic feet per second (cfs) flow of water per 1,000 MW installed capacity is needed for modern power plants. In addition, some 13 cfs of flow

5. See U.S. Executive Office of the President [109] for a review of conventional siting practices.

is required as makeup for cooling tower blowdown requirements. An emergency cooling supply may also be required for nuclear plants. The total amount of water required depends upon the generation mix and the cooling system employed. When considering energy parks the problem increases by orders of magnitude, which naturally restricts both the availability of sites and the types of cooling systems that can be employed. Evaporation loss for a 1,000 MW nuclear plant using cooling towers is about 2.5 to 3 percent. When using deep cooling ponds the loss can be reduced to around 1 percent. Hence, the makeup required for a nominally rated 10GW energy park would be about 250 cfs, slightly less for cooling ponds, slightly more natural draft towers. A cooling pond for a 10GW park would, at about 1.5 acreas per MW installed capacity, need to be 15,000–20,000 acres (about 25 square miles).

Other hydrologic factors that must be considered in power plant siting include flood protection and water quality. The site must be out of flood range of the water source and the source itself must be suitable with respect to stream flow, temperature, stratification and depth.

Geology and topography. Areas of excessive relief are unsuitable for power plant construction since extensive excavation would be required. In addition, the condition of underlying geologic structure is crucial, particularly for nuclear plants. Foundation is a major capital item in power plant construction. Geologic stability falls into the class of problems that are often referred to as 'common mode' failures which become more serious when considering energy parks since a single geologic disturbance could cause multiple accidents.

Meteorology. Air quality is a prime limiting meteorological factor in siting fossil power plants. Air basins, where the probability of inversion is high, are avoided. In an all-fossil energy park, ambient air quality standards could present an overwhelming obstacle. This, along with the fact that nuclear plants require excessive amounts of cooling water, presents a favorable case for the hybrid park alternative, which would require less cooling water than an NEC of equivalent capacity and would also much more easily meet ambient air quality standards. Meteorology is also a critical factor in the type of cooling employed. Many energy park sites (of those recently proposed) could not meteorologically support 100 percent utilization of evaporative cooling towers since they might occasionally cause excessive fogging and icing in the local area.

Transport access. In conventional dispersely sited power plants existing transport facilities are extemely important in siting. In general, highways would provide access for plant construction and operation and rail or water access would be required for deliver of heavy equipment and/or fuels. The larger plants become the more sophisticated transportation facilities must become, but costs of constructing these facilities become overshadowed by the other capital costs involved in energy parks. Construction of high quality roads and long rail spurs become economically justified when considering energy parks. The Pennsylvania study [85] proposes a 50 percent fossil and 50 percent nuclear energy park (12GW$_e$). In this case, since the fossil fuel that will be employed in the future is almost certain to be coal, such a park would require, at a 70 percent load factor, some 30,000 tons of coal per day (300 'Big-John Hopper' railroad cars). A month's stockpile would then be 900,000 to a million tons of coal. Simple logistics of handling such volumes of coal become a non-trivial problem. However, some economies of scale could be achieved in fuel handling facilities.

Land use. Depending upon the cooling method employed the total land requirement of an energy park could vary greatly. The General Electric energy park study [68] estimates a 60 percent reduction in total land requirements for an energy park versus equivalent dispersed generation (NEC-26GW$_e$ using cooling towers and excluding the transmission right-of-way requirement). If one includes other cooling sources such as ponds and suitable transmission corridors, this relative reduction in land requirement would be considerably reduced. However, the total cost, as a percentage of total investment, would probably be less than one percent. When considering nuclear plants one must also include the regulatory requirements of the exclusion and low population zones. However, with hybrid parks, ash and fuel storage facilities, as well as cooling ponds, could overlap with these zones. Isard et al. [71] argue that the exclusion and low population zones of energy parks should be extended significantly which would increase the total land requirement accordingly.

2. Plant design considerations

Standardized plant construction. The GE study [68] generically investigates potential economies of scale in power plant construction due to modular design which might become feasible with energy parks. Special on-site metal and concrete fabrication facilities could also contribute to construction economies of scale. A large number of units would justify use of large bridge cranes in handling equipment and structures in plant construction. The NRC [110] cites the use of bridge cranes as the source of largest potential savings in construction costs. They predict a total saving of

12 percent; General Electric [68] and the Pennsylvania study [85] both predict slightly larger savings.

Transmission. Few recent energy park studies consider transmission a technical difficulty, it is considered, rather, an economic one. In general, an energy park would imply the need for longer transmission lines, increased stability preserving equipment and redundancy (in terms of both independent transmission corridors and systems). The NRC [110] estimates that the total transmission penalty (construction costs and line loss) would be several percent of the total cost of the park with current technology. The Transmission Task Force of the Pennsylvania study indicates that high voltage DC transmission would become economically competitive at distances over 300 miles to load centers (in Pennsylvania) due to reduced right-of-way requirements. On a site-specific basis, for small energy parks, total transmission cost could reduce or negate the otherwise overall economic scale advantage the park may have over an equivalent amount of dispersed generation.

Safety and reliability. The growing controversy over nuclear power plant safety has cast a shadow of uncertainty on the heretofore optimistic future of the nuclear power industry in this country. The NRC argues that the accident risks in dispersed nuclear plants are very low (Rasmussen [88]) and claim further for most accident classes there would be little difference in risk between NEC's and dispersely sited nuclear plants. No attempt will be made here to support or refute these contentions since the controversy concerning nuclear safety is still very much an issue. However, the public perception of the issue, which will be discussed shortly, is very important to overall acceptability of energy parks. Indeed, the acceptance of dispersed nuclear power plants is probably prerequisite to acceptance of the energy park concept unless radical designs (e.g. floating NEC's) prove significantly safer than an equivalent amount of dispersely sited nuclear generation.

4.3. ISSUES OF PRACTICALITY

The principal technical issues involved in planning energy parks have been generally resolved for many potential sties, i.e. it is clear that they can be built in some form, but questions remain as to whether or not they should be built. These questions fall into the realm of *practicality* (a term coined by the NRC which includes issues concerning ownership-management, socioeconomic impact, government involvement, jurisdiction [antitrust], financing, licensing, interstate cooperation and public acceptance). Foremost among these, as far as generic feasibility of the energy park concept is concerned are those of financing, government organization and regulation. Table 4.2 shows the projected capital costs of various size energy parks (including escalation and interest on construction) interspersed with the GNP's of countries and net sales of selected companies for the year 1973. This table dramatically shows the level of capital investment required for some alternatives of energy park development currently being considered. Of principal concern then is the source of capital for such an enterprise. Many feel that government involvement would be required to lessen this financial burden on both the utilities and the capital market. The GE study flatly declares [68]:

Unless there is change in the current outlook, government involvement, especially at the Federal level, will almost certainly be required to support the private sector. Governmental incentives could be offered in the form of facilitating favorable interest rates, use of the power of eminent domain, and the planning assistance of a Federal energy agency....

Financing is only one facet of proposed government involvement. Existing regulatory requirements would need extensive revision to accommodate the energy park concept. Boundaries of states and service areas do not conform well to energy park development. In nearly all cases proposed energy parks would require the cooperative effort of a number of utilities. In Pennsylvania, parks were initially proposed to service the entire Pennsylvania-New Jersey-Maryland (PJM) interconnection which includes utilities from all of those states. However, state policy required that, for an energy park sited in Pennsylvania, partial ownership or control by out-of-state utilities could not be permitted.

The GE study [68] proposes establishment of 'interstate compacts' that would handle control and regulation of energy parks. The study proposed that these 'compacts' might be initially financed by the federal government. They would also be used as a vehicle to administer the complex coordination of the diverse views in energy facility siting, such as establishment of a 'designated site' licensing procedure. Such a procedure would abolish the traditional plan of site selection which is carried out by the utility concerned subject to government approval. The designated site concept would require site selection to be the joint responsibility of utilities, government and appropriate planning authorities. Under the current procedure, the extended judicial and regulatory review period (often referred to as 'regulatory lag') appears inadequate in terms of

TABLE 4.2. Gross national products of nations, net sales of companies, and projected costs of energy park investments.

Country, company, or energy park	GNP, net sales, or total investment (billions of dollars)
U.S.	1155.2
U.S.S.R.	549.0
Japan	335.2
West Germany	285.7
France	217.8
United Kingdom	151.5
People's Republic of China	141.0
Italy	122.1
Canada	102.6
India	58.3
Poland	56.2
Spain	50.9
Netherlands	50.5
Australia	50.5
Brazil	49.6
46 to 48 GW$_e$ nuclear energy center (U.S.)	48.0
East Germany	46.5
Sweden	43.6
Mexico	40.7
Belgium	39.2
Czechoslovakia	37.9
Switzerland	37.0
General Motors Corporation	35.8
Romania	31.4
Argentina	27.2
EXXON Corporation	25.7
Yugoslavia	24.0
American Telephone & Telegraph Corporation	23.5
24 GW$_e$ nuclear energy center	23.2
Denmark	23.2
Ford Motor Corporation	23.0
Austria	22.7
South Africa	21.1
Royal Dutch Shell Group (Corporation)	18.7
20 GW$_e$ energy park (hybrid design)	18.5
Hungary	18.3
Iran	17.4
Turkey	17.0

Source: (GNP and corporation data): Congressional Research Service (October 1974).

duplication of effort and, in some cases, inefficient since the basic issues of siting priorities are not effectively aired.

The question of the potential regional impact of an energy park, which is a central issue of the present work, is perhaps the most important non-technical question regarding energy parks on a site-specific basis. The issues of regional impact can perhaps be posed in another light – that of 'fairness' in the allocation of costs and benefits. For instance, tax benefits that might accrue through energy park development would, under current practices, be distributed directly to the county in which the park is sited (usually indirectly through the state government) and not to the surrounding region, which would experience a significant impact as well. The proper allocation of such tax revenues must be tailored to the likely impact on the surrounding region. The degree of environmental quality degradation a region would tolerate in the process of providing power to a distant load center must also be considered.

The major economic impacts of energy park development on employment and demands for regional goods and services would, in most cases, be beneficial. The park development would entail a long-term stable work force. A typical regional economy is a hierarchy of urban centers, ranging from small villages to sprawling metropolitan centers. A new enterprise the size of an energy park would more than likely alter this regional hierarchy, perhaps even raising a second-class urban center to a point of regional economic dominance. However, if the region is viewed in the aggregate and if interindustry structure (not necessarily the scale of operation) is assumed to remain constant, then impacts can be investigated apart from these locational aspects. This assumption will allow us to use the policy programming procedure of Part One to investigate regional aspects of energy park development.

In addition to investigating the likely impacts of an energy park on a particular region one must also be concerned with the public perception of those impacts. Utility spokesmen are constantly concerned about public naiveté in technical aspects of electrical energy production. Usually, however, the public dismisses the educational efforts of utilities as biased propaganda. The involvement of a presumably unbiased governmental or planning agency in education of technical aspects of energy parks would be necessary to gauge properly realistic rather than misinformed public opinion concerning potential energy park development in the region. Clearly, public reaction to proposed energy park development is the most uncertain issue surrounding overall feasibility of energy parks.

Finally there is one last issue regarding the ultimate construction and operation of an energy park. It is possible that a cooperative arrangement made between several utilities in an energy park project might constitute an unreasonable restraint of trade, in violation of existing antitrust laws (a lengthy discussion of this issue is given in the GE Study [68]). This problem could occur with the energy park either as a buyer of construction materials or as a seller of its product. Existing antitrust review by the NRC may not be sufficient to deal with all aspects of the problem.

4.4. THE ENERGY PARK CONCEPT AS A SYSTEM

The energy park concept as it has been unfolded in the previous sections presents an extremely complex, large-scale planning problem that might have significant far-reaching and diverse impacts upon the region in which energy park is sited. The magnitude and diversity of these impacts depend primarily upon the size of the proposed park and the interests in the region that might be affected by that park. Hence, the system affected by energy park development can quite naturally be described by the generalized *policy-making system* formulated in chapter 2. In that formulation we recall that the *policy-making interests* are those which would be affected by energy park development.

The response of these interests to potential impacts of a proposed energy park development plan, or at least their perception of those impacts, would profoundly affect the success of implementing the development plan. Each of the policy-making interests would exert his influence in order to insure that his own preferences were realized or, in the systems terminology of chapter 2, each policy-maker (a purposeful subsystem) acts according to his own goals and objectives. The entire system is driven from state to state by the collective actions of all the subsystems. The actions are carried out through system variables over which policy makers have varying amounts of influence.

An example of this might be the following: in a region where an energy park is proposed, industrial producers could offer favorable prices to promote energy park development since park construction promotes a long-term market for their products. Environmentalists, on the other hand, might exert their influence to block energy park development in the region since operation of the park is likely to cause pollution levels that are unacceptable to them.

The overall system that the energy park will affect is, of course, adaptive, since the actions of policy makers may change in response to actions of other policy makers. This adaption takes place by means of feedback between subsystems. This interaction and feedback among subsystems then is ultimately of principal concern in planning for energy park development in a region. However the first step is to account for the preferences and relative influence of policy makers and then project an outcome that corresponds to these preferences and influences but that is also consistent with the technical relationships controlling the overall system operation. In chapter 3 we referred to these relationships as *system process functions.*

To account for all the policy-making interests, relevant system variables, process functions and measures of influence and preference in the energy park problem would be impossible. We cannot hope to capture all aspects of the energy park planning problem. However, as mentioned earlier, the issues of practicality are currently of most concern in assessing the overall feasibility of energy parks. Many of these practicality issues are concerned with regional economic, environmental, energy, and employment impacts. Hence, in the specific evaluation of alternative energy park development plans we shall focus our analysis on these impacts.

In Part One, the policy programming framework was constructed to account for a number of *policy-making interests* in long-range planning problems such as energy park development. Many of the likely economic, environmental, and employment impacts of development plans could also be found by means of the model of system process functions developed in chapter 2. As we recall, the policy programming framework serves to link the hierarchy of objectives of policy-making interests to this model of system process functions.

We see then that the system which will be considered in planning for energy park development is limited to effects that such a development would have on interindustry activity and system variables closely related to that activity, such as energy consumption, pollution emission, and employment. Such a system, we remember from chapter 2, is called an *energy-environment system.* We realize that this system is not the complete picture that must be considered in energy park development for a region but it does capture a critical portion of the complete picture. Many of the other considerations, such as political ones concerning allocation of tax revenues or antitrust, are not included here.

The regional energy-environment system. For purposes of this study the collection of important policy-making interests in energy park development planning was taken from those identified by a research team at the University of Pennsylvania Energy Center (Denton et al. [6]) concerned with a systems approach to planning for regional assessment. The collection of policy makers identified in that study were viewed as those most important in determining the future of a regional energy-environment system in the context of a number of alternative decision environments, one of which was planning for energy park development. Table 4.3 summarizes this collection of policy-making interests.

In the application of policy programming to investigate energy park development in a target region (chapter 5) this collection of policy-making interests will be aggregated where possible into the most important actor groups, aligning similar objectives of these policy makers. This will be done in order to reduce the scale of computation.

Also, for purposes of this study, a specific model of system process functions was developed for the target region which is presented in the next chapter.

In summary, whereas this chapter provided some insight into the energy park concept, the next chapter will address development of an energy park at a specific site and assemble the information needed to employ the policy programming procedure for analyzing alternative energy park planning options for this site.

TABLE 4.3. Classification of actors in energy-environment systems.

Government		Non-government	
Legislative	· Federal · State	Energy supplier	· Coal · Oil

Executive	· President
	· Governors
	· Federal agencies:
	  EPA, NRC,
	  ERDA, FPC,
	  FEA, etc.
	· State Agencies:
	  DER, PUC
			· Gas
			· Nuclear
			· Alternative sources:
			  LNG
			  SNG
			  Shale
			  Solar

		Electric utility Equipment manufacturers	
Judiciary	· Federal · State	Heavy construction Individual consumer Industrial consumer Environmentalist Labor leader Financier	

5. The energy park planning region

5.1. INTRODUCTION

A number of studies have considered the selection of potential sites for energy parks (see e.g. NUREG-0001[110], Gilbert Associates [69], U.S. Nuclear Regulatory Commission [111], and Blair et al. [61]). All of these studies have screened potential sites in terms of technical feasibility alone. The most suitable site in these studies is taken to be the one that best satisfies the technical requirements of an energy park design.

Gilbert Associates [69] has identified a number of prime siting regions for energy park development in Pennsylvania (see fig. 5.1). One of

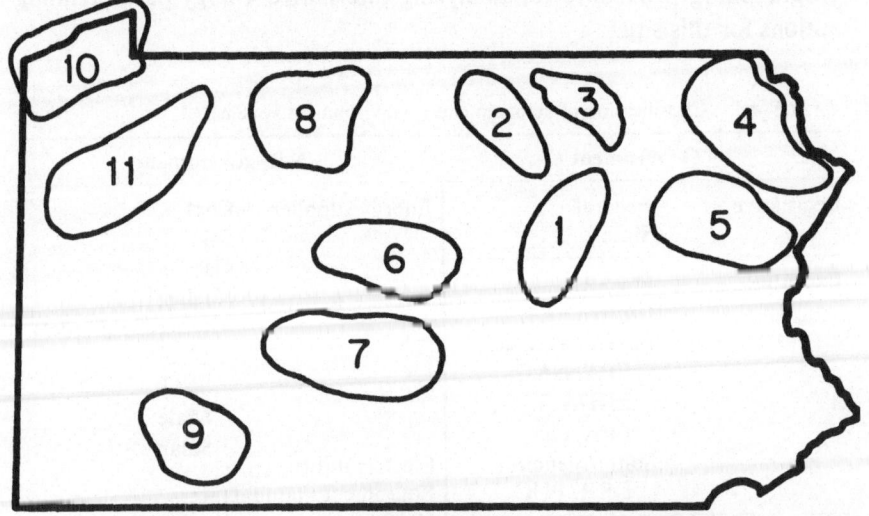

1. Montour	5. Tobyhanna	9. Donegal
2. Cherry Flats	6. Pine Glen	10. Erie
3. Leraysville	7. Chestnut Ridge	11. Sheakleyville
4. Ararat	8. Clermont	

FIG. 5.1. Energy park siting regions in Pennsylvania.

the most promising of these regions in terms of technical considerations such as availability of cooling water, geologic stability, transportation access, etc, is the Montour siting region (Region 1 in fig. 5.1).

The Montour area was chosen as the planning region in this study for two reasons: (1) there is a sufficient amount of information available to construct a suitable model of system process functions for the region, and (2) the siting region is technically flexible enough to support a number of alternative energy park designs, ranging from NEC's employing cooling towers which require a relatively small land commitment, to hybrid energy parks using cooling ponds which have a large land requirement.

The second of these reasons allows one to consider an energy park design at the site that may be technically sub-optimal but better suited to the goals and objectives of policy-making interests in the siting region.[1] For instance, from the electric utility's point of veiw, a large NEC may be technically feasible and preferred at the Montour site. However, a significant amount of coal-fired generation in the park design may be better suited to regional industries as these industries may be able to supply more of the required construction materials for a hybrid park than for an NEC.

In the following sections, the characteristics of the specific site proposed for energy park development in the planning region will be sketched. Then the existing economic structure, patterns of energy consumption, pollution emission, and employment of the region will be characterized in terms of a model of system process functions. This model will be a generalized input-output model of the form presented in chapter 2.

The input-output model will then be used to construct a *reference scenario*, as defined in chapter 3, for the region. Similarly, the model will provide a number of *future scenarios* corresponding to various energy park development plans. We recall, from chapter 3, that the *reference* and *future scenarios* must be *consistent scenarios*. In this case, consistency will be assured since the model of system process functions, which we recall is used to define the conditions of consistency, is also used to construct these *future scenarios*.

The reference and future scenarios form the basic data needed to invoke the policy programming procedure which will be used to investigate alternative planning options for energy park development in the Montour region. This application of policy programming will be a subject of chapter 6.

---

1. The Pennsylvania Energy Park Development Group [85] has included a requirement in their energy park siting criteria of at least 50 percent coal generation in all park designs even though electric utilities would generally rather build nuclear plants.

5.2. THE MONTOUR REGION

The Montour site considered here is located about 20 miles NNE of the junction of the west and north branches of the Susquehanna River. The site is three miles north of Danville (Montour's county seat) which is located on the north branch of the Susquehanna River, the proposed cooling source. The siting area actually straddles both Montour and Columbia counties with the bulk of the proposed generation being constructed in a large flat valley in the Derry and Madison townships of the respective counties (the northern half of the site). The northern edge of the site is bounded by the Pennsylvania railroad. Interstate Highway 80 is several miles south of the site and Pennsylvania Route 54 borders the western edge of the park. The southern and eastern portions of the site are hilly areas with natural creek bed catchments where a number of ponds and a pumped storage facility were eventually sited. The station area, the northern and central portions of the site, are wooded and cultivated hills of low relief. This area has an average elevation of 550 feet, while the pond area ranges up to over 1000 feet. The entire siting area's location with respect to the major transportation arteries in the two county area is shown in fig. 5.2. The site is discussed in detail in Blair et al. [61] and Gilbert Associates [69].

Completion of the site selection and layout of any large energy facility must, in general, be accomplished in conjunction with the final choice of the generation mix. Site flexibility or limitations would shape choices concerning the final list of projects and schedule for construction of these projects in an every park design. The Montour site is flexible enough to facilitate technically a variety of energy park designs.

In the policy programming framework, the alternative designs of energy parks might be posed as a number of *future scenarios*. Before composing these scenarios, we first characterize the region in terms of its existing economic structure and patterns of energy consumption, employment, and pollution emission which, as we recall from Part One, constitute a *reference scenario* if the structure and patterns are *consistent* with the model of system process functions. If these data are not consistent with the system model, the model should be altered until consistency is reached.

From this reference scenario we might easily project one future scenario: that of *not* developing an energy park in the region. The costs and benefits of other future scenarios must be balanced against this 'reference-extended' future scenario in order to evaluate whether or not energy park development in any form is acceptable in the target region.

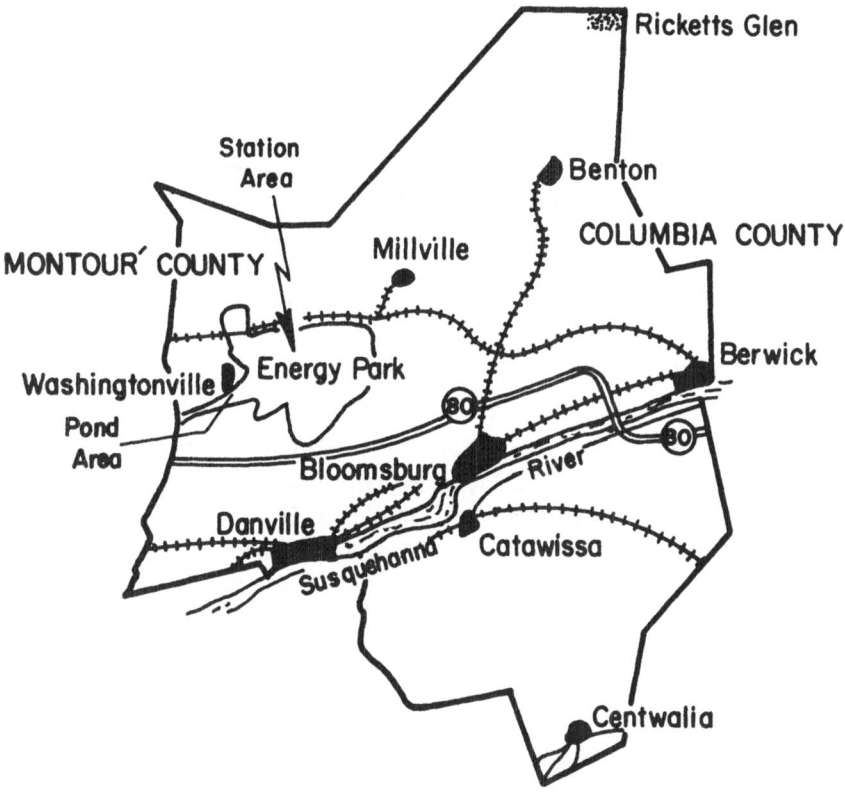

FIG. 5.2. Energy park siting area.

In using the policy programming procedure we recall that the following must be performed: (1) construction of a *composite scenario* that is formed from the alternative future scenarios, the relative preference policy-making interests have for these future scenarios and the influence each policy maker has in ensuring that his preferences are realized; and (2) ensurance of the *consistency* (*Definition 2*, chapter 3) of the composite scenario with the relevant *system process functions*.

In the following section the existing economic (interindustry) structure of the Montour region will be examined and the *reference scenario* for that region will then be composed. In section 5.4 the *future scenarios* will then be composed. Construction of the *composite scenario* will be accomplished in the next chapter.

5.3. INTERINDUSTRY STRUCTURE

As mentioned, the economic structure of the planning region can be conveniently represented as an input-output table. Actually, no input-output table exists for the planning region but several input-output studies have been completed for counties in north central Pennsylvania, several of which have characteristics similar to those of the planning region.[2] Hence, we might attempt to modify one of these existing tables to develop an input-output table that more realistically represents the interindustry structure of the actual siting region. The input-output table for Clinton County (Gamble and Raphael [133]) was obtained for this purpose.

Modification techniques are often employed in input-output studies where primary data collection (survey) is impractical (see Morrison and Smith [155]). A number of these non-survey modification techniques have appeared in recent years. (see McMenamin and Haring [151]). Perhaps the most contemporary of these approaches, and perhaps the best suited for this study in terms of data requirements (which will be discussed shortly), is the biproportional or RAS Method. The mechanics of this method are discussed in some detail in Appendix B. Briefly, the method requires estimates of intermediate inputs and intermediate and total outputs in addition to the base input-output table. The RAS procedure systematically modifies the base table until it conforms to the data given for the target region: given (1) U, the observed vector of intermediate outputs; V, the observed vector of intermediate inputs; and X, the vector of total outputs for the target region, and (2) A, the matrix of technical coefficients for the base input-output table, the RAS procedure finds A' such that:

$$A'X = U$$
$$iA'\hat{X} = V$$

where $i = [1, 1, \ldots, 1]$ and $\hat{X} = \begin{bmatrix} x_1 \ldots 0 \\ \cdot\ x_2 \quad \cdot \\ \cdot \quad\quad \cdot \quad \cdot \\ \cdot \quad\quad\ \cdot \quad \cdot \\ 0 \ldots \quad x_n \end{bmatrix}$

This new matrix of technical coefficients (A') is assumed to describe the input-output structure of the target region.

2. Pennsylvania State University has constructed a set of economic input-output tables for a number of Pennsylvania counties (see Gamble and Raphael [133]).

The modified table. The RAS technique was used to modify the Clinton County table to represent one for the planning region. We recall that in order to use the RAS modification procedure, the U, V and X vectors for the target region must somehow be estimated or observed, which is a much simpler task than actual survey-based construction of a complete inter-industry transactions table. In practice, it is often easier to estimate a region's value-added and final demand vectors than the vectors of total intermediate inputs and outputs. The vector U can simply be found, as was shown in chapter 2, as the difference between total outputs and final demands. Similarly, V can be found as the difference between total outputs and the value-added vectors.

In this study estimates of value-added and total outputs were taken from county industry data for the region (Pennsylvania Department of Commerce [158]). Final demand estimates were broken into the following four categories: (1) personal consumption expenditures, (2) capital formation, (3) government expenditures, and (4) net exports. The last three categories were projected from the county historical trends but the first, personal consumption, was estimated from personal consumption expenditure profiles and projected levels of regional income. Tables 5.1 and 5.2 give the projections of the total output, value-added and final demand used in the RAS projection of the Clinton County table. Note the level of industry detail (aggregation) is dictated by the size of the base input-output table. Much larger tables exist (e.g. Philadelphia's 500-sector table – see Isard and Langford [143]) but these regions are neither geographically close nor compatible with the siting region since they usually describe an urban rather than a rural region. Moreover, use of such a large table is well beyond the scope of the present study.

The reliability of using the RAS technique to modify input-output tables has been discussed extensively elsewhere (see Malizia and Bond [150], Hewings [137], Schafer [161], and Blair [125]). Briefly, RAS is most reliable when technological patterns between the base and target regions are not radically different. This is the rationale for choosing the base table to be one of a region that is geographically close to the target region. Note that in the present application, RAS was used to both regionalize and update (in time) the base table.

In the process of determining the reference scenario we are constructing a model of system process functions for the region. The RAS procedure gives us part of this model – the A matrix of economic input-output coefficients.

In chapter 3 the extension of input-output to a generalized framework

TABLE 5.1. Projected regional data for RAS estimation*

Sector classification	Total final demand	Total value-added	Total sector output
1. Mining	7.938	6.080	11.249
2. Food processing	12.296	6.730	25.091
3. Textiles	68.073	28.835	98.575
4. Matr. processing	152.336	94.567	239.859
5. Prtg. publish	1.687	2.195	4.417
6. Fabr., assem.	156.087	94.543	220.075
7. Chemicals	37.943	20.628	55.896
8. Sawmills	0	.254	.677
9. Pulpwood	1.062	.492	1.331
10. Agr. feeds, fert.	1.844	1.958	4.028
11. Misc. industry	7.313	4.264	11.026
12. Agriculture	8.881	7.226	18.907
13. Education	9.499	2.194	10.967
14. Foodstores	6.259	38.204	57.304
15. Gas stations	3.563	15.236	27.771
16. Auto dealers	15.014	44.218	64.448
17. Clothing	4.598	7.110	19.219
18. Bars, restaurants	2.687	8.437	16.704
19. Jwlry., drgs.	2.473	6.883	9.783
20. Furn. and appl.	3.509	8.070	19.731
21. Hrdware, bld. mat.	3.009	16.897	23.436
22. Depart., variety	.727	21.606	31.611
23. Farm equipment	2.312	2.526	7.232
24. All other retail	5.974	7.167	11.939
25. Hotels, motels	3.098	4.354	5.866
26. Enter., recreation	1.705	1.526	2.933
27. Finance	5.449	10.553	14.577
28. Rl. estate insur.	1.777	15.704	19.643
29. Laundry	1.973	.896	2.515
30. Prof. pers. svcs.	5.706	9.145	13.687
31. Nonprof. pers. svcs.	5.652	4.045	12.598
32. Transportation	13.993	23.791	40.547
33. Construction	6.735	13.315	30.174
34. Wholesale	20.480	25.524	71.351
35. Electric, gas	7.710	9.120	15.755
36. Phone	1.348	7.336	9.111
37. Water, tv, radio	1.298	2.695	5.430
38. Nonprofit	7.978	8.897	32.365
39. Local govt.	2.827	3.923	7.221
40. Public schools	14.563	7.318	24.987
41. County govt.	1.487	.415	4.034
42. State govt.	5.633	1.774	8.410
43. Federal govt.	2.227	0	2.815
44. Labor	137.101	89.063	337.228
45. Rent	6.918	4.276	16.739
46. Transfer	31.631	13.167	43.586
47. Proprietary	6.063	12.348	48.557
48. Overflow	4.188	0	16.091

* in millions of dollars.

TABLE 5.2. Components of final demand*

Sector classification	Total external	Capital formation	State government	Federal government
1. Mining	0.00	7938.77	0.00	0.00
2. Food processing	607.24	11689.37	0.00	0.00
3. Textiles	125.02	67948.37	0.00	0.00
4. Matr. processing	0.00	152336.87	0.00	0.00
5. Prtg. publish	0.00	1687.77	0.00	0.00
6. Fabr., assem.	0.00	156087.47	0.00	0.00
7. Chemicals	0.00	37943.57	0.00	0.00
8. Sawmills	0.00	0.00	0.00	0.00
9. Pulpwood	0.00	1062.67	0.00	0.00
10. Agr. feeds, fert.	0.00	1687.77	0.00	157.17
11. Misc. industry	0.00	7313.67	0.00	0.00
12. Agriculture	0.00	7938.77	0.00	943.01
13. Education	321.48	3563.07	5222.26	392.92
14. Foodstores	3321.96	2937.97	0.00	0.00
15. Gas stations	625.10	2937.97	0.00	0.00
16. Auto dealers	785.84	14189.77	0.00	39.29
17. Clothing	1660.98	2937.97	0.00	0.00
18. Bars, restaurants	1000.16	1687.77	0.00	0.00
19. Jwlry., drgs.	160.74	2312.87	0.00	0.00
20. Furn. and appl.	571.52	2937.97	0.00	0.00
21. Hrdware, bld. mat.	71.44	2937.97	0.00	0.00
22. Depart., variety	500.08	4188.17	0.00	39.29
23. Farm equipment	0.00	2312.87	0.00	0.00
24. All other retail	535.80	5438.37	0.00	0.00
25. Hotels, motels	160.74	2937.97	0.00	0.00
26. Enter., recreation	642.96	1062.67	0.00	0.00
27. Finance	303.62	1687.77	0.00	3457.70
28. Rl. estate insur.	89.30	1687.77	0.00	0.00
29. Laundry	285.76	1687.77	0.00	0.00
30. Prof. pers. svcs.	2143.20	3563.07	0.00	0.00
31. Nonprof.pers.svcs.	214.32	5438.37	0.00	0.00
32. Transportation	428.64	13564.67	0.00	0.00
33. Construction	410.78	5438.37	60.72	825.13
34. Wholesale	0.00	20440.77	0.00	39.29
35. Electric, gas	357.20	7313.67	0.00	39.29
36. Phone	285.76	1062.67	0.00	0.00
37. Water, tv, radio	196.46	1062.67	0.00	39.29
38. Nonprofit	893.00	6688.57	121.45	275.04
39. Local govt.	428.64	1062.67	1335.93	0.00
40. Public schools	0.00	1687.77	12600.23	275.04
41. County govt.	0.00	1062.67	425.07	0.00
42. State govt.	107.16	1062.67	4463.21	0.00
43. Federal govt.	0.00	1062.67	850.14	314.34
44. Labor	0.00	30442.37	60.72	2003.89
45. Rent	625.10	1062.67	5191.90	39.29
46. Transfer	0.00	1062.67	0.00	30569.18
47. Proprietary	0.00	6063.47	0.00	0.00
48. Overflow	0.00	4188.17	0.00	0.00

* in thousands of dollars

was discussed in some detail. We recall that a *total impact matrix* can be given by:

$$D = \begin{bmatrix} u \\ \hline v \\ \hline w \end{bmatrix}$$

or: $T^* = \begin{bmatrix} (I - A)^{-1} \\ \hline D^* \end{bmatrix}$

u = direct energy coefficients ($m_1 \times n$)

v = direct pollution coefficients ($m_2 \times n$)

w = direct employment coefficients ($1 \times n$)

m_1 = number of energy types

m_2 = number of pollutant types

Hence, the *total* impact due to some final demand presented to the input-output economy can be found by $\mathcal{X} = T^*Y$, where $\mathcal{X}$ is the vector of industry outputs, energy consumption, pollution emission and employment which were collectively defined in chapter 3 as a *scenario* $\mathcal{X} = (X, X^*) = (X, u^*, v^*, w^*)$. T^* then describes a complete model of system process functions for the region.

Data for u, v and w for this study were derived from Herendeen [134], Toscas [164], and Bezdek and Hannon [124], respectively. Appendix A discusses how the matrices u, v and w were developed from the available data of these sources. The only attempt to regionalize these data was to adjust the resulting coefficients (u, v and w) so that their use in the base projection, which is the multiplication of D by the X, would yield regional total pollution, energy consumption and employment estimates as control totals:

$$\mathcal{C} = D\,X, \quad \text{where} \quad D = \begin{bmatrix} u \\ \hline v \\ \hline w \end{bmatrix}$$

and where $\mathcal{C}$ is the vector of control totals of energy consumption, air pollution and employment for the target region and X_r is the reference level of total industry outputs used in the RAS estimation. The final coefficients u, v and w are shown in tables 5.3 and 5.4. Note that in table 5.4 only air pollution has been included as in indicator of environmental quality. Given more data, other indicators could have been included. Note also that the control totals form the basis for altering T^* to insure that the *reference scenario* is *consistent* with the model of system process functions.

Reference scenario. Given the reference levels of final demand (those used

TABLE 5.3. Direct energy and employment coefficients*

Sector classification	Coal	Crude oil & gas	Refined petroleum	Electricity	Gas utilities	Employment
1. Mining	0.00472	0.04424	0.00498	0.00039	0.00410	18.9
2. Food processing	0.00259	0.	0.00252	0.00162	0.01644	14.6
3. Textiles	0.00358	0.	0.00231	0.00391	0.02256	26.3
4. Matr. processing	0.00115	0.	0.00068	0.00223	0.01322	25.2
5. Prtg. publish	0.00010	0.	0.00104	0.00076	0.00490	36.6
6. Fabr., assem.	0.00106	0.	0.00197	0.00166	0.00980	26.8
7. Chemicals	0.02551	0.00717	0.03470	0.00495	0.01300	14.7
8. Sawmills	0.00090	0.	0.00683	0.00059	0.00562	36.5
9. Pulpwood	0.02356	0.	0.01502	0.00407	0.02612	20.3
10. Agr.feeds, fert.	0.00038	0.	0.07793	0.00075	0.00404	7.9
11. Misc. industry	0.01170	0.06912	0.00909	0.00257	0.01876	22.7
12. Agriculture	0.00041	0.	0.00195	0.00045	0.00264	50.3
13. Education	0.	0.	0.	0.00115	0.01139	51.7
14. Foodstores	0.	0.	0.	0.00026	0.00294	25.5
15. Gas stations	0.	0.	0.00537	0.00021	0.00239	54.6
16. Auto dealers	0.00045	0.	0.00025	0.00033	0.00204	53.2
17. Clothing	0.00059	0.	0.00142	0.00091	0.00928	46.6
18. Bars, restaurants	0.	0.	0.	0.00044	0.00501	27.7
19. Jwlry., drgs.	0.00101	0.	0.00047	0.00126	0.00952	25.5
20. Furn. and appl.	0.00122	0.	0.00086	0.00114	0.01025	25.5
21. Hrdware, bld. mat.	0.	0.	0.	0.00029	0.00329	24.7
22. Depart., variety	0.	0.	0.	0.00041	0.00469	25.5
23. Farm equipment	0.00396	0.	0.00168	0.00162	0.01620	18.2
24. All other retail	0.	0.	0.	0.	0.	48.8
25. Hotels, motels	0.	0.	0.00179	0.00088	0.00919	36.4
26. Enter., recreation	0.	0.	0.00179	0.00214	0.02432	40.8
27. Finance	0.	0.	0.00074	0.00006	0.00069	36.0
28. Rl. estate insur.	0.	0.	0.	0.00017	0.00196	24.7
29. Laundry	0.	0.	0.	0.00245	0.02791	61.2
30. Prof.pers.svcs.	0.	0.	0.00567	0.00011	0.00120	51.0
31. Nonprof.pers.svcs.	0.	0.	0.	0.00053	0.00605	24.0
32. Transportation	0.00036	0.	0.06868	0.00024	0.00279	39.2
33. Construction	0.	0.	0.01748	0.00010	0.00095	28.0
34. Wholesale	0.	0.	0.06857	0.00027	0.00305	8.2
35. Electric, gas	0.39495	0.90306	0.04753	0.00152	0.00335	16.7
36. Phone	0.	0.	0.00289	0.00039	0.00381	32.1
37. Water, tv, radio	0.	0.	0.00038	0.00155	0.01536	29.1
38. Nonprofit	0.	0.	0.	0.00030	0.00339	25.5
39. Local govt.	0.	0.	0.01825	0.00041	0.00063	32.0
40. Public schools	0.	0.	0.	0.00116	0.01151	34.2
41. County govt.	0.	0.	0.	0.00027	0.00313	33.5
42. State govt.	0.	0.	0.	0.00011	0.00124	47.3
43. Federal govt.	0.	0.	0.00055	0.00091	0.01347	48.9
44. Labor	0.	0.	0.	0.	0.	45.9
45. Rent	0.	0.	0.00107	0.00041	0.00469	0.0
46. Transfer	0.	0.	0.	0.	0.	0.0
47. Proprietary	0.	0.	0.	0.	0.	0.0
48. Overflow	0.	0.	0.	0.	0.	0.0

*Energy coefficients are in 10^6 btu/per dollar of output; employment is measured in man-yrs. per million dollars of output.

TABLE 5.4. Direct air pollution coefficients

	Part	HC	SO$_2$	CO	NO$_x$
			(tons/10^6 \$)		
Food processing	23.25	0.07	6.65	0.20	0.22
Textiles	15.10	0.11	10.37	0.29	2.93
Matr. processing	223.58	0.49	40.47	1.38	15.01
°Prntg. publish	0.61	0.01	0.75	0.01	28.48
Fabr., assem.	3.44	0.03	3.21	0.07	1.16
Chemicals	206.53	58.24	139.36	42.16	0.45
Sawmills	2.94	0.03	3.76	0.06	1.17
Pulpwood	2.94	0.03	3.76	0.06	1.17
Agr. feeds, fert.	3.43	0.03	2.42	0.07	0.76
Electric, gas	0.00	0.00	609.00	0.00	0.00

in the RAS estimation) we can characterize the *reference scenario* formally by:

$$\mathcal{X}_r = \mathbf{T}^* \mathbf{Y}_r$$

where $\mathbf{Y}_r$ is the reference vector of final demands, and $\mathcal{X}_r$ is the *reference scenario* given by $\mathcal{X} = (\mathbf{X}, \mathbf{X}^*)$.

The corresponding industry transactions might also be included in this reference scenario so that one could observe the flow of economic goods in the region. Development or future scenarios might alter this transaction's picture significantly. The transactions for the reference scenario are easily found by:

$$\mathbf{Z} = \mathbf{A}\,\hat{\mathbf{X}} \text{ where } \hat{\mathbf{X}} = \begin{bmatrix} x_1 \dots 0 \\ \cdot\ x_2\ \cdot \\ \cdot \\ \cdot \quad \cdot \\ \cdot \quad \cdot \\ 0 \dots x_n \end{bmatrix}$$

The reference final demands and total dollar outputs are given in table 5.5. The corresponding $\mathbf{X}^*$ values are also given in that table.

TABLE 5.5. Reference scenario*

Industry sector	Y	X	State variables	X*
1. Mining	7.94	11.25	Coal	8.93
2. Food processing	12.30	25.09	Crude oil and gas	15.89
3. Textiles	68.07	98.58	Refined petroleum	12.85
4. Matr. processing	152.34	239.86	Electric utilities	1.97
5. Prtg. publish	1.69	4.42	Natural gas utilities	11.96
6. Fabr., assem.	156.09	229.08	Part emissions	68073.99
7. Chemicals	37.94	56.00	HC emissions	3397.83
8. Sawmills	0.00	0.68	SO_2 emissions	28050.58
9. Pulpwood	1.06	1.33	CO emissions	2741.60
10. Agr. feeds, fert.	1.84	4.03	NO_x emissions	4315.21
11. Misc. industry	7.31	11.08	Regional employ-	
12. Agriculture	8.88	18.91	ment	53146.17
13. Education	9.50	10.97		
14. Foodstores	6.26	57.30		
15. Gas stations	3.56	27.77		
16. Auto dealers	15.01	64.45		
17. Clothing	4.60	19.22		
18. Bars, restaurants	2.69	16.70		
19. Jwlry., drgs.	2.47	9.78		
20. Furn. and appl.	3.51	19.73		
21. Hrdware, bld. mat.	3.01	23.44		
22. Depart., variety	4.73	31.61		
23. Farm equipment	2.31	7.23		
24. All other retail	5.97	11.94		
25. Hotels, motels	3.10	5.87		
26. Enter., recreation	1.71	2.93		
27. Finance	5.45	14.58		
28. Rl. estate insur.	1.78	19.64		
29. Laundry	1.97	2.51		
30. Prof. pers. svcs.	5.71	13.09		
31. Nonprof.pers.svcs.	5.65	12.60		
32. Transportation	13.99	40.55		
33. Construction	6.74	30.17		
34. Wholesale	20.48	71.35		
35. Electric, gas	7.71	15.76		
36. Phone	1.35	9.11		
37. Water, tv, radio	1.30	5.43		
38. Nonprofit	7.98	32.37		
39. Local govt	2.83	7.22		
40. Public schools	14.56	24.97		
41. County govt.	1.49	4.03		
42. State govt.	5.63	8.41		
43. Federal govt.	2.23	2.82		
44. Labor	137.10	337.23		
45. Rent	6.92	16.74		
46. Transfer	31.63	43.59		
47. Proprietary	6.06	48.56		
48. Overflow	4.19	16.09		

* Industry output and final demand are measured in millions of dollars; energy consumption in 10^{15} btu and pollution emissions in tons.

5.4. FUTURE SCENARIOS

We now develop a set of *future scenarios*, each of which corresponds to an alternative plan for energy park development in the Montour region. One future scenario, however, we recall from the introduction to this chapter, is that of not developing an energy park in the region at all.

Scenario 1: reference scenario extended. Without energy park development of any kind one might expect the region to follow its historical pattern. Such projections (in an aggregate sense) are made routinely by the Pennsylvania Department of Commerce in their county industry reports [158]. Assuming no radical changes in regional technology (or purchasing patterns) the future projection of industrial output can be estimated by taking the estimates of final demand from these reports and then using input-output analysis to compute corresponding industry outputs. With the total impact coefficients matrix (T^*), we can project the levels of energy consumption, environmental pollution (air quality), and employment as well. All of these projections are made assuming no change in technology or regional purchasing patterns. The final demands were computed according to consumption profiles of households in the target region (see Isard et al. [71]).

Scenarios 2 to 5: development scenarios. Energy park development would have two primary economic impacts upon the surrounding region: (1) income and employment affects resulting from primary employment in energy park construction and secondary employment for support services and industry; and (2) impact of purchases of material for plant construction. An important feature in using input-output projections for construction of future scenarios is that imports can be easily taken into account. This is important since not all plant construction materials would be regionally produced. The generalized input-output framework also provides other important state variables consistent with the economic projection (i.e. energy consumption, total emissions, employment).

   In addition to the 'reference extended' future scenario, five other planning scenarios will be included:

2. Nuclear energy center ($10GW_e$);
3. Hybrid energy park ($10GW_e$) (50 percent nuclear, 50 percent fossil);
4. Nuclear energy center ($20GW_e$);
5. Hybrid energy park ($20GW_e$) (50 percent nuclear, 50 percent fossil);
6. Hybrid energy park ($10GW_e$) (including planned use of site flexibility).

Several studies have presented the manpower and material requirements for nuclear and fossil fired power plants (e.g. Rombough and Koen[91], Reiner et al. [89], ERDA-3[168], and NUREG-0001[110]). Results from these studies will be used in predicting the impact of energy park development on the Montour region.

Construction materials impact. Among those components included in the list of material requirements are some specialized high-cost items that are not (and are not likely to be in the future) produced in the region. Isard et al. [71] identified these components as the following:

> reactor vessels
> control rod drive systems
> reactor coolant pumps
> steam generators
> pressurizers
> compressors (waste gas)
> drumming stations
> major cranes
> diesel generators.

In an urban region, e.g. Philadelphia, most other items could be regionally produced in quantities that would not saturate local production capacity. However, in regions such as Montour, many more construction items would be externally produced. This could be due to the fact that the regional economy does not produce the item, but not necessarily. Alternatively, the geographic advantage of regional firms (lower transportation costs) could be offset by other interregional cost factors (see Reiner et al. [89]), since all items are procured through competitive bidding.

The best estimates of material requirements for a nuclear power plant are found in NRC's WASH-1230, volume 1 [101] which was subsequently updated in WASH-1345 [105]. The latter report includes the incremental changes in material costs that are reflected by regulatory changes between 1971 and 1974. Fossil plant estimates were taken from WASH-1230. In extending these requirements to energy park designs some economies of scale will be realized. WASH-1345 summarizes some of the facilities that would not be duplicated for more than one nuclear unit:

1. most site preparation costs;
2. roads and railroad spurs;

3. in the case of a standardized plant, engineering and architectural costs, environmental impact statement preparation, etc;
4. administrative facilities;
5. miscellaneous facilities: cooling water discharge, warehouses, control rooms, laboratory facilities, fences, guardhouses, radiation monitoring equipment, waste disposal facilities, fuel handling and storage facility, computer equipment.

Isard et al. [71] estimate that a 17 percent reduction in material costs would be experienced with the additional unit in a two-unit plant. United Engineers [99] estimate a 28 percent reduction in construction costs of a $48GW_e$ NEC due to economies of scale. The NECSS [108] estimates a 12 percent reduction for a four unit ($\sim 4.8GW_e$). Using these estimates, the material requirements (treated as a final demand presented to the regional economy) for scenarios 2, 3, 4 and 5 were then computed and adjusted for apparent economies of scale and then again for the capacity of regional industry to supply these requirements. The rest of the material requirements would be purchased externally. The estimates of material requirements for the future scenarios (2–5) are given in table 5.6. These estimates, adjusted for regional capacity, are given in table 5.7. (see Isard et al. [71]).

Income impact (wages and salaries). Some fraction of the income paid to construction workers in an energy park project would be spent in the region and, as a result, would provide a direct and indirect impact on the regional economy. Isard et al. [71] estimate a leakage factor of 30 percent for the Philadelphia region (wages and salaries paid to construction workers). Gilbert Associates [69] estimate a 50 percent leakage factor for more rural regions, of which the present siting region is one. It is assumed here that 50 percent of the employment income will be spent in the region. Fifty percent may be an optimistic figure where one assumes that availability of services within the region (water, sewer, schools, etc) will prompt significant migration. The Pennsylvania State University [87] presents less optimistic figures. Total income of wages and salaries for a single nuclear plant ($2400MW_e$) has been estimated at $220.5 million [71]. With the 50 percent leakage factor just assumed, the income spent by energy park generated wages and salaries would be about $45/kw ($110.13 million for a $2400MW_e$ plant). Actually, the work force at an energy park site would be less than that required of the equivalent generation capacity installed at dispersed sites. The NRC [110] and General Electric [68] have both made estimates of work force requirements for energy parks. At $25,000 per man-year, a figure used in Isard et al. [71], the total income in wages and salaries for a

TABLE 5.6. Energy park material requirements

Material cost category	millions of dollars Coal 1GW	BWR 1GW	PWR 1GW	Coal 5GW	BWR 5GW	PWR 5GW	Coal 10GW	BWR 10GW	PWR 10GW	HYB* 10GW	PWR 20GW	HYB* 20GW
Land	1.00	1.00	1.00	4.36	4.36	4.36	4.36	8.40	8.40	8.40	15.95	16.80
Structures	10.54	14.30	12.32	45.99	62.32	53.72	88.61	120.05	103.49	99.71	196.47	192.10
Boiler plant equipment	0.00	40.17	45.08	0.00	175.12	196.56	0.00	337.35	378.69	196.56	718.90	378.69
Turbine equipment	38.53	0.00	0.00	168.06	0.00	0.00	323.81	0.00	0.00	168.06	0.00	323.81
Electric plant equipment	36.07	46.90	46.43	157.30	204.41	202.41	303.08	393.78	389.96	358.70	740.28	693.03
Miscellaneous plant equipment	7.19	7.69	8.29	31.37	33.52	36.13	60.44	64.58	69.62	67.50	132.15	130.05
Professional services	1.86	2.10	2.13	9.29	10.49	10.67	18.58	20.98	21.34	19.96	42.68	39.92
Other undistributed costs	13.65	23.75	23.75	59.53	103.52	103.55	114.70	199.43	199.49	163.08	378.71	314.20
Temporary facilities	1.05	1.26	1.26	5.25	6.30	6.30	10.50	12.60	12.60	11.55	25.20	23.10
Construction equipment	5.52	6.43	6.03	24.10	28.05	26.31	46.43	54.04	50.69	50.41	96.23	97.12
Construction services	1.75	2.07	1.95	8.78	10.32	9.82	17.55	20.65	19.65	18.60	39.30	37.20
Administrative materials	2.60	4.20	4.20	13.00	21.00	21.00	26.00	42.00	42.00	34.00	84.00	68.00
Total material cost	119.78	149.87	152.46	527.01	659.41	670.83	1018.10	1273.86	1295.93	1197.84	2489.88	2314.03

* HYB indicates hybrid energy park (50% coal generation and 50% nuclear generation).

5 GW columns include 12% economies of scale.
10 GW columns include 15% economies of scale.
20 GW columns include 19% economies of scale.

TABLE 5.7.   Energy park material requirements adjusted for regional production capacity*

Material cost category	NEC(PWR) 10GW	HYB 10GW	NEC(PWR) 20GW	HYB 20GW $(10^6\$)$
Land	8.40	8.72	15.95	16.80
Structures	21.28	20.55	39.93	39.28
Reactor	77.87	40.50	146.10	77.43
Boiler plant equipment	0.00	34.63	0.00	66.21
Turbine equipment	80.19	74.12	150.45	141.71
Electric plant equipment	69.62	67.50	132.16	130.05
Miscellaneous plant equipment	21.34	19.96	42.68	39.92
Professional services	41.02	33.60	76.96	64.24
Other undistributed costs				
Temporary facilities	12.60	11.55	25.20	23.10
Construction equipment	10.42	10.39	10.56	19.86
Construction services	4.04	3.83	7.99	7.61
Administrative materials	42.00	34.00	84.00	68.00
Total material cost	388.8	359.4	741.0	694.2

* Fifty percent leakage factor.

$10\,GW_e$ and $20\,GW_e$ energy park would be about \$857 million and \$1714 million, respectively, over the construction period of the park. In order to determine the corresponding values of final demand for goods and services presented to the regional economy by this income, the estimated total values of income were allocated to final demand sectors according to regional consumption trends. These allocated final demands are given in table 5.8. Note that these data have also been adjusted for relevant consumer margins (transportation, wholesale and retail). The base data for these adjustments were taken from the *Survey of Current Business* [140] and are summarized in table 5.9.

In order to complete the projections of the future scenarios we need to compute the impact of these projected final demands (materials and income) via the total impact matrix. These projections then take on, of course, the assumptions of any input-output projection: fixed input requirements and constant returns to scale. Table 5.10 shows the total final demands presented to the regional economy by the four planning scenarios (2, 3, 4, and 5) which

TABLE 5.8. Allocated final demands due to energy park wage and salary income for each projected future scenario (10^3 $).

Sector classification	2.(PWR-10GW)	3.(HYB-10GW)	4.(PWR-20GW)	5.(HYB-20GW
1. Mining	0.267	0.250	0.456	0.437
2. Food processing	31212.702	29170.750	53236.619	51048.812
3. Textiles	2716.730	2539.000	4633.675	4443.250
4. Matr. processing	3233.005	3021.500	5514.237	5287.625
5. Prntg, publish	614.715	574.500	1048.482	1005.375
6. Fabr., assem.	8196.467	7660.250	13979.956	13405.437
7. Chemicals	14.445	13.500	24.637	23.625
8. Sawmills	704.862	658.750	1202.219	1152.812
9. Pulpwood	152.742	142.750	260.519	249.812
10. Agr. feeds, fert.	123.050	115.000	209.875	201.250
11. Misc. industry	11373.565	10629.500	19398.837	18601.625
12. Agriculture	4902.205	4581.500	8361.237	8017.625
13. Education	1228.895	1148.500	2096.012	2009.875
14. Foodstores	8045.597	7519.250	13722.631	13158.687
15. Gas stations	1810.975	1692.500	3088.812	2961.875
16. Auto dealers	1978.182	1848.750	3373.969	3235.312
17. Clothing	12401.567	11590.250	21152.206	20282.937
18. Bars, restaurants	13125.155	12266.500	22386.362	21466.375
19. Jwlry., drgs.	4644.335	4340.500	7921.412	7595.875
20. Furn. and appl.	4132.607	3862.250	7048.606	6758.937
21. Hrdware, bld.mat.	2150.700	2010.000	3668.260	3517.500
22. Depart, variety	8834.722	8256.750	15068.569	14449.312
23. Farm equipment	524.032	489.750	893.794	857.062
24. All other retail	0.000	0.000	0.000	0.000
25. Hotels, motels	11624.480	10864.000	19826.800	19012.000
26. Enter, recreation	866.700	810.000	1478.250	1417.500
27. Finance	10714.980	10014.000	18275.550	17524.500
28. Rl. estate ins.	9841.860	9198.000	16786.350	16096.500
29. Laundry	0.000	0.000	0.000	0.000
30. Prof. pers. svcs.	8403.245	7853.500	14332.637	13743.625
31. Nonprof.pers.svcs.	3739.382	3494.750	6377.919	6115.812
32. Transporttation	2252.617	2105.250	3842.081	3684.187
33. Construction	29.692	27.750	50.644	48.562
34. Wholesale	9366.245	8753.500	15975.137	15318.625
35. Electric, gas	3156.232	2949.750	5383.294	5162.062
36. Phone	3000.547	2804.250	5117.756	4907.437
37. Water, tv, radio	572.450	535.000	976.375	936.250
38. Nonprofit	10901.695	10186.500	18594.013	17829.875
39. Local govt.	11113.287	10386.250	18954.906	18175.937
40. Public schools	1077.490	1007.000	1837.775	1762.250
41. County govt.	0.000	0.000	0.000	0.000
42. State govt.	470.800	440.000	803.000	770.000
43. Federal govt.	3852.000	3600.000	6570.000	6300.000
44. Labor	0.000	0.000	0.000	0.000
45. Rent	53995.410	50463.000	92094.975	88310.250
46. Transfer	0.000	0.000	0.000	0.000
47. Proprietary	0.000	0.000	0.000	0.000
48. Overflow	399.377	373.250	681.181	653.187
Totals	267500.000	250000.000	452500.000	437500.000
Total man-years $25K/man-year	21400.000	20000.000	36500.000	35000.000

TABLE 5.9. Base data for removal of margins*

BEA sector	Producers' prices	Transportation costs	Wholesale & retail trade margins	Purchasers' prices
01	13.4	1.3	1.3	16
02	7	1	4	12
03	449	85	496	1029
05	25.4	1.3	0.3	27
06	67.4	3.3	0.3	71
07	36	24	1	61
08	257	3	0	260
09	4	1	0	5
10	31	8	0	39
13	6431.6	95.7	98.7	6626
14	448.4	1.3	8.3	458
15	189	1	68	258
16	73.4	1.3	4.3	79
17	89	3	10	102
18	96.4	1.3	5.3	103
19	323	5	34	362
20	22	1	2	25
21	21	1	1	23
22	165	7	10	182
23	1632	59	415	2106
24	55.4	3.3	8.3	65
25	29	1	2	32
26	49	4	11	64
27	466	11	11	488
28	65	1	1	67
29	71	1	7	79
30	52	2	31	85
31	138.6	9.7	31.7	180
32	29.6	0.7	10.7	41
33	5	0	0	5
34	21	2	8	31
35	8	1	1	10
36	23	1	3	27
37	286	16	21	323
38	34	1	1	36
39	10.6	0.7	0.7	12
40	933.4	22.3	40.3	996
41	337	8	52	397
42	107.6	1.7	4.7	114
43	735	6	19	760
44	2942	76	1051	4069
45	2428	39	314	2781
46	1107.6	24.7	68.7	1191
47	3461	35	158	3654
48	3204.6	30.7	180.7	3416
49	1888	38	171	2097
50	98	1	7	106
51	3352	17	335	3704
52	1729	37	282	2048
53	2886	42	170	3098
54	127	2	19	148
55	67	1	7	75
56	2936.6	15.7	85.7	3038
57	616.6	1.7	25.7	644
58	209.6	3.7	67.7	281
59	9054	264	1372'	10690
60	809.6	0.7	3.7	814
61	3430	35	367	3832
62	1131.6	9.7	213.7	1355
63	920	4	302	1226
64	532.6	12.7	80.7	626
82	99	2	13	114
83	1286.5	0	3197.5	4484

* Prices are in 1963 dollars.

TABLE 5.10. Total final demands for future scenarios

| Sector classification | Future scenarios (millions of dollars) | | | |
	NEC 10GW[2]	Hybrid 10GW[3]	NEC 20GW[4]	Hybrid 20GW[5]
1. Mining	0.000	0.000	0.000	0.000
2. Food processing	31.213	29.171	53.237	51.049
3. Textiles	2.717	2.539	4.634	4.443
4. Matr. processing	3.233	3.021	5.514	5.288
5. Prtg. publish	0.615	0.574	1.048	1.005
6. Fabr., assem.	29.478	28.206	53.907	52.685
7. Chemicals	0.014	0.013	0.025	0.024
8. Sawmills	0.705	0.659	1.202	1.153
9. Pulpwood	0.153	0.143	0.261	0.250
10. Agr. feeds, fert.	0.123	0.115	0.210	0.201
11. Misc. industry	249.470	237.769	467.659	453.858
12. Agriculture	4.902	4.581	8.361	8.018
13. Education	1.229	1.148	2.096	2.010
14. Foodstores	8.046	7.519	13.723	13.159
15. Gas stations	1.811	1.692	3.089	2.962
16. Auto dealers	1.978	1.849	3.374	3.235
17. Clothing	12.402	11.590	21.152	20.283
18. Bars, restaurants	13.125	12.266	22.386	21.466
19. Jwlry., drgs.	4.644	4.340	7.921	7.596
20. Furn. and appl.	4.133	3.862	7.049	6.759
21. Hrdware, bld.mat.	2.151	2.010	3.668	3.517
22. Depart., variety	8.835	8.257	15.069	14.449
23. Farm equipment	0.524	0.490	0.894	0.857
24. All other retail	0.000	0.000	0.000	0.000
25. Hotels, motels	11.624	10.864	19.827	19.012
26. Enter., recreation	0.867	0.810	1.478	1.417
27. Finance	10.715	10.014	18.276	17.524
28. Rl. estate insur.	18.242	17.919	32.732	32.899
29. Laundry	0.000	0.000	0.000	0.000
30. Prof. pers. svcs.	95.465	79.290	183.284	153.594
31. Nonprof.pers.svcs.	3.739	3.495	6.378	6.116
32. Transportation	19.325	18.073	37.986	35.620
33. Construction	12.630	11.578	25.251	23.149
34. Wholesale	9.366	8.753	15.975	15.319
35. Electric, gas	3.156	2.950	5.383	5.162
36. Phone	7.269	6.796	13.654	12.891
37. Water, tv, radio	0.572	0.535	0.976	0.936
38. Nonprofit	10.902	10.188	18.594	17.830
39. Local govt.	11.113	10.386	18.955	18.176
40. Public schools	1.077	1.007	1.838	1.762
41. County govt.	0.000	0.000	0.000	0.000
42. State govt.	0.471	0.440	0.803	0.770
43. Federal govt.	3.852	3.600	6.570	6.300
44. Labor	0.000	0.000	0.000	0.000
45. Rent	53.995	50.463	92.095	88.310
46. Transfer	0.000	0.000	0.000	0.000
47. Proprietary	0.000	0.000	0.000	0.000
48. Overflow	0.399	0.373	0.681	0.653

TABLE 5.11. Future scenario total outputs*

Sector classification	(2)	(3)	(4)	(5) (10^6 $)
1. Mining	3.851	3.631	7.185	6.885
2. Food processing	41.203	38.483	70.941	67.909
3. Textiles	8.428	7.869	14.596	13.958
4. Matr. processing	15.848	14.841	27.883	26.709
5. Prtg. publish	3.885	3.646	6.949	6.663
6. Fabr., assem.	68.636	64.756	122.605	118.271
7. Chemicals	5.404	5.044	9.393	8.976
8. Sawmills	1.267	1.188	2.260	2.160
9. Pulpwood	0.170	0.159	0.292	0.280
10. Agr. feeds, fert.	2.087	1.948	3.613	3.456
11. Misc. industry	250.590	238.825	469.661	455.788
12. Agriculture	17.454	16.302	30.147	28.848
13. Education	2.284	2.131	3.994	3.813
14. Foodstores	47.118	43.914	83.688	79.665
15. Gas stations	22.643	21.093	40.764	38.722
16. Auto dealers	45.804	42.745	82.682	78.671
17. Clothing	23.070	21.527	40.329	38.503
18. Bars, restaurants	20.436	19.070	35.513	33.929
19. Jwlry., drgs.	10.317	9.576	18.181	17.242
20. Furn. and appl.	16.575	15.454	29.429	28.023
21. Hrdware, bld. mat.	21.184	19.868	38.383	36.699
22. Depart., variety	28.512	26.604	50.493	48.136
23. Darm equipment	4.822	4.501	8.593	8.182
24. All other retail	6.405	5.979	11.237	10.728
25. Hotels, motels	12.633	11.812	21.634	20.746
26. Enter., recreation	1.555	1.453	2.715	2.597
27. Finance	18.970	17.731	33.165	31.724
28. Rl. estate insur.	39.430	37.790	71.435	69.893
29. Laundry	0.383	0.354	0.685	0.644
30. Prof. pers. svcs.	101.712	85.095	194.518	164.241
31. Nonprof. pers. svcs.	9.419	8.802	16.653	15.906
32. Transportation	48.587	45.873	92.520	88.375
33. Construction	30.940	28.675	58.260	54.577
34. Wholesale	48.536	45.335	86.737	82.697
35. Electric, gas	8.138	7.633	14.421	13.829
36. Phone	15.857	14.854	29.388	27.934
37. Water, tv, radio	3.831	3.580	6.777	6.471
38. Nonprofit	28.732	26.634	50.784	48.072
39. Local govt.	14.557	13.590	25.134	24.041
40. Public schools	9.196	8.575	16.555	15.747
41. County govt.	1.514	1.407	2.721	2.578
42. State govt.	2.832	2.644	4.984	4.757
43. Federal govt.	4.511	4.221	7.777	7.459
44. Labor	177.210	119.848	331.983	222.869
45. Rent	63.760	59.574	109.646	105.016
46. Transfer	5.959	5.585	10.508	10.073
47. Proprietary	75.622	69.396	139.485	130.102
48. Overflow	4.937	4.641	8.755	8.417
49. Coal	6.645	6.276	12.068	11.634
50. Crude oil and gas	24.879	23.597	45.871	44.362
51. Refined petroleum	11.751	10.996	21.661	20.609
52. Electric utilities	1.205	1.138	2.199	2.119
53. Natural gas utilities	9.274	8.748	16.899	16.263
54. Part. emissions	5994.325	5608.916	10489.340	10044.803
55. HC emissions	328.639	306.829	571.440	546.197
56. SO_2 emissions	6945.868	6509.985	12259.915	11752.624
57. CO emissions	265.448	247.912	461.943	441.665
58. NO_x emissions	467.266	438.271	826.413	792.670
59. Regional employment	38314.004	35479.642	69904.075	65888.358

* Industry output and final demand are measured in millions of dollars; energy consumption in 10^{15} btu and pollution emissions in tons.

include both the material requirements (adjusted for regional capacity) and the total wages and salaries (adjusted for estimated leakage). Table 5.11 gives the corresponding total outputs required to support these final demands computed from the total impact matrix ($\mathcal{X} = \mathbf{T}^*\mathbf{Y}$).

Scenario 6: hybrid planning scenario (optimized). One alternative for energy park development in the siting region not yet discussed is the park design proposed in Blair et al. [61] which was designed to make maximum use of the surrounding site. For instance, in this proposal cooling ponds were designed to be used as primary cooling sources for some plants, emergency cooling for nuclear plants, pumped storage and recreation. Existing fossil facilities and transmission corridors were expanded rather than having entirely new facilities built. The major construction projects in this alternative are somewhat different from those of the other development scenarios. The projects required in this alternative are listed in table 5.12. The corresponding material requirements are shown in table 5.13.

The corresponding final demands (including construction wages and salaries) for development scenario 6 are given in table 5.14 which, as before, include the costs adjusted for regional capacity and leakage. Finally, table 5.15 summarizes all the projected future scenarios, including the extended reference (scenario 1).

TABLE 5.12. Major construction projects (scenario [6])

Project description	Rating	Cooling method
Coal-fired Unit (2)	1400 MW	2 towers
1. Coal-fired Unit	800 MW	1 tower
2. Coal-fired Unit	800 MW	1 tower
3. Nuclear unit (2)	2100 MW	2 towers
4. Nuclear unit (2)	2100 MW	2 towers
5. Nuclear unit (2)	2100 MW	pond
6. Nuclear unit	1050 MW	pond
7. Pumped storage plant	1200 MW	−
8. Combustion turbines	200 MW	−
Total installed generation	11750 MW	(8 towers)
	(10350 MW new)	(6 towers new)
9. Pumped storage dams (3)		
10. Cooling pond dam		
11. Sludge pond dam		
12. Management facility		
13. Substations (2)		
14. Railroad spur, access roads		

TABLE 5.13.   Energy park material costs (scenario[6])

Cost category	(Millions of dollars) Coal 2GW$_e$	PWR 7GW$_e$	Hybrid[a] 10GW$_e$	Hybrid[b] 10GW$_e$
Land	1.79	6.03	7.82	7.82
Structures	18.84	74.28	845.38	392.23
Reactor	0.00	271.82	271.82	126.12
Boiler plant equipment	68.86	0.00	68.86	31.95
Turbine equipment	64.45	279.90	453.66	210.48
Electric plant equipment	12.85	49.97	62.82	62.82
Miscellaneous plant equipment	3.72	14.94	18.65	18.65
Professional services	24.39	143.19	167.59	77.75
Temporary facilities	2.10	8.82	10.92	10.92
Construction equipment	9.87	36.39	46.26	21.46
Construction services	3.51	13.75	17.26	8.01
Administrative materials	5.20	29.40	34.60	34.60
Total material costs				

a   Including economies to scale and special equipment: dams, combustion turbines, hydro-
turbines.
b   Adjusted for regional capacity.

Summary We note that since we used the model of system process functions to construct the *future scenarios*, they are *consistent* according to *Definition 2* of chapter 3. Hence, in this chapter, we have characterized the target region and posed a number of alternative consistent future scenarios of energy park development. In addition, a 'reference-extended' future scenario was constructed which did not include an energy park development plan. The scenarios were quantified in terms of a set of system variables by using input-output analysis (a model of the system process functions) to compute the likely impacts of the various development plans.

We next consider the application of the policy programming framework to the problem of planning for energy park development in the Montour region. In this application we shall use the scenarios just developed.

TABLE 5.14. Future scenario [6]*

Industry classification	X (10^6 $)	X,·X*
1. Mining	0.000	5.838
2. Food processing	29.171	9.722
3. Textiles	2.539	3.078
4. Matr. processing	3.021	29.637
5. Prtg. publish	0.574	3.992
6. Gabr., assem.	399.891	420.826
7. Chemicals	0.013	2.497
8. Sawmills	0.659	0.704
9. Pulpwood	0.143	0.033
10. Agr. feeds, fert.	0.115	0.760
11. Misc. industry	463.464	459.571
12. Agriculture	4.581	5.261
13. Education	1.148	1.372
14. Foodstores	7.519	46.902
15. Gas stations	1.692	26.203
16. Auto dealers	1.849	55.217
17. Clothing	11.590	13.615
18. Bars, restaurants	12.267	11.992
19. Jwlry., drgs.	4.340	7.304
20. Furn. and appl.	3.862	16.067
21. Hrdware, bld. mat.	2.010	28.258
22. Depart., variety	8.257	24.683
23. Farm equipment	0.490	3.608
24. All other retail	0.000	4.678
25. Hotels, motels	10.864	4.490
26. Enter., recreation	0.810	1.605
27. Finance	10.014	10.055
28. Rl. estate insur.	17.014	36.787
29. Laundry	0.000	0.526
30. Prof. pers. svcs.	128.219	127.725
31. Nonprof. pers. svcs.	3.495	8.186
32. Transportation	17.028	66.647
33. Construction	10.948	34.705
34. Wholesale	8.753	53.233
35. Electric, gas	2.950	8.623
36. Phone	6.535	19.011
37. Water, tv, radio	0.535	3.310
38. Nonprofit	10.188	24.549
39. Local govt.	10.386	4.570
40. Public schools	1.007	12.969
41. County govt.	0.000	2.458
42. State govt.	0.440	2.082
43. Federal govt.	3.600	1.012
44. Labor	0.000	277.655
45. Rent	50.463	10.872
46. Transfer	0.000	14.121
47. Proprietary	0.000	98.223
48. Overflow	0.373	20.998
49. Coal		9.493
50. Crude oil and gas		39.829
51. Refined petroleum		15.588
52. Electric utilities		2.226
53. Natural gas utilities		15.702
54. Part. emissions		8868.113
55. SO_2 emissions		173.634
56. CO emissions		8254.522
57. NO_x emissions		178.671
58. Regional employment		1058.604
Total		59477.541

* Industry output and final demand are measured in millions of dollars; energy consumption in 10^{15} btu and pollution emissions in tons.

TABLE 5.15. Future scenarios*

Industry classification (System variable)	1.Reference Ext.		2.Small NEC-10GWe		3.Small Hyb-10GWe		4.Large NEC-20GWe		5.Large Hyb-20GWe		6.Hybrid-Opt.	
	Y	X,X*	Y	X,X*	Y	X,X*	Y	X,X*	Y	X,X*	Y	X,X*
1. Mining	2.224	3.151	0.000	3.851	0.000	3.631	0.000	7.185	0.000	6.885	0.000	5.838
2. Food processing	3.444	7.028	31.213	41.203	29.171	38.483	53.237	70.941	51.049	67.909	29.171	9.722
3. Textiles	19.067	27.611	2.717	8.428	2.539	7.869	4.634	14.596	4.443	13.958	2.539	3.078
4. Matr. processing	42.670	67.185	3.233	15.848	3.021	14.841	5.514	27.883	5.288	26.709	3.021	29.637
5. Prntg., publish	0.473	1.237	0.615	3.885	0.574	3.646	1.048	6.949	1.005	6.663	0.574	3.992
6. Fabr., assem.	43.720	64.164	29.478	68.636	28.206	64.756	53.907	122.605	52.685	118.271	399.891	420.826
7. Chemicals	10.628	15.685	0.014	5.404	0.013	5.044	0.025	9.393	0.024	8.976	0.013	2.497
8. Sawmills	0.000	0.190	0.705	1.267	0.659	1.188	1.202	2.260	1.153	2.160	0.659	0.704
9. Pulpwood	0.298	0.373	0.153	0.170	0.143	0.159	0.261	0.292	0.250	0.280	0.143	0.033
10. Agr. feeds, fert.	0.517	1.128	0.123	2.087	0.115	1.948	0.210	3.613	0.201	3.456	0.115	0.760
11. Misc. industry	2.049	3.088	249.470	250.590	237.769	238.825	467.659	469.661	453.858	455.788	463.464	459.571
12. Agriculture	2.488	5.296	4.902	17.454	4.581	16.302	8.361	30.147	8.018	28.848	4.581	5.261
13. Education	2.661	3.072	1.229	2.284	1.148	2.131	2.096	3.994	2.010	3.813	1.148	1.372
14. Foodstores	1.753	16.051	8.046	47.118	7.519	43.914	13.723	83.688	13.159	79.665	7.519	46.902
15. Gas stations	0.998	7.779	1.811	22.643	1.692	21.093	3.089	40.764	2.962	38.722	1.692	26.203
16. Auto dealers	4.206	15.052	1.978	45.804	1.849	42.745	3.374	82.682	3.235	78.671	1.849	55.217
17. Clothing	1.288	5.383	12.402	23.070	11.590	21.527	21.152	40.329	20.283	38.503	11.590	13.615
18. Bars, restaurants	0.753	4.579	13.125	20.436	12.266	19.070	22.386	35.513	21.466	33.929	12.267	11.992
19. Jwlry, drgs.	0.693	2.740	4.644	10.317	4.340	9.576	7.921	18.181	7.596	17.242	4.340	7.304
20. Furn. and appl.	0.983	5.527	4.133	16.575	3.862	15.454	7.049	29.429	6.759	28.023	3.862	16.067
21. Hrdware, bld. mat.	0.843	6.065	2.151	21.184	2.010	19.868	3.668	38.383	3.517	36.699	2.010	28.258
22. Depart., variety	1.324	8.554	8.835	28.512	8.257	26.604	15.069	50.493	14.449	48.136	8.257	24.683
23. Farm equipment	0.648	2.126	0.524	4.822	0.490	4.501	0.894	8.593	0.857	8.182	0.490	3.608
24. All other retail	1.673	3.344	0.000	6.405	0.000	5.979	0.000	11.237	0.000	10.728	0.000	4.678
25. Hotels, motels	0.868	1.643	11.624	12.633	10.864	11.812	19.827	21.634	19.012	20.746	10.864	4.490
26. Enter., recreation	0.478	0.822	0.867	1.555	0.810	1.453	1.478	2.715	1.417	2.597	0.810	1.605
27. Finance	1.526	4.033	10.715	18.970	10.014	17.731	18.276	33.165	17.524	31.724	10.014	10.055
28. Rl. estate ins.	0.498	5.502	18.242	39.430	17.919	37.790	32.732	71.435	32.899	69.893	17.014	36.787

	C1	C2	C3	C4	C5	C6	C7	C8	C9	C10	C11	C12
29. Laundry	0.553	0.704	0.000	0.383	0.000	0.354	0.000	0.685	0.000	0.644	0.000	0.526
30. Prof. pers. svcs.	1.598	3.666	95.465	101.712	79290	85.095	183.284	194.518	153.594	164.241	128.219	127.725
31. Nonprof. pers. svcs.	1.583	3.529	3.739	9.419	3.495	8.802	6.378	16.653	6.116	15.906	3.495	8.186
32. Transportation	3.920	11.357	19.325	48.587	18.073	45.873	37.986	92.520	35.620	88.375	17.028	66.647
33. Construction	1.886	8.452	12.630	30.940	11.578	28.675	25.251	58.260	23.149	54.577	10.948	34.705
34. Wholesale	5.737	19.986	9.366	48.536	8.753	45.335	15.975	86.737	15.319	82.697	8.753	53.233
35. Electric, gas	2.160	4.413	3.156	8.138	2.950	7.633	5.383	14.421	5.162	13.829	2.950	8.623
36. Phone	0.378	2.552	7.269	15.857	6.796	14.854	13.654	29.388	12.891	27.934	6.535	19.011
37. Water, tv, radio	0.364	1.521	0.572	3.831	0.535	3.580	0.976	6.777	0.936	6.471	0.535	3.310
38. Nonprofit	2.235	9.066	10.902	28.732	10.188	26.634	18.594	50.784	17.830	48.072	10.188	24.549
39. Local govt.	0.792	2.023	11.113	14.557	10.386	13.590	18.955	25.134	18.176	24.041	10.386	4.570
40. Public schools	4.079	6.994	1.077	9.196	1.007	8.575	1.838	16.555	1.762	15.747	1.007	12.969
41. County govt.	0.417	1.130	0.000	1.514	0.000	1.407	0.000	2.721	0.000	2.578	0.000	2.458
42. State govt.	1.578	2.356	0.471	2.832	0.440	2.644	0.803	4.984	0.770	4.757	0.440	2.082
43. Federal govt.	0.624	0.789	3.852	4.511	3.600	4.221	6.570	7.777	6.300	7.458	3.600	1.012
44. Labor	38.402	94.458	0.000	127.210	0.000	119.848	0.000	231.985	0.000	222.869	0.000	277.655
45. Rent	1.938	4.689	53.995	63.760	50.463	59.574	92.095	109.646	88.310	105.016	50.463	10.872
46. Transfer	8.860	12.209	0.000	5.959	0.000	5.585	0.000	10.508	0.000	10.073	0.000	14.121
47. Proprietary	1.698	13.601	0.000	75.622	0.000	69.396	0.000	139.485	0.000	130.102	0.000	98.223
48. Overflow	1.173	4.507	0.399	4.937	0.373	4.641	0.681	8.755	0.653	8.417	0.373	20.998
49. Coal	2.501		6.645		6.276		12.068		11.634		9.493	
50. Crude oil and gas	4.451		24.879		23.597		45.871		44.362		39.829	
51. Refined petroleum	3.598		11.751		10.996		21.661		20.609		15.588	
52. Electric utilities	0.551		1.205		1.138		2.199		2.119		2.226	
53. Natural gas utilities	3.349		9.274		8.748		16.899		16.263		15.702	
54. Part. emissions	19067.525		5994.325		5608.916		10489.340		10044.803		8868.113	
55. HC emissions	951.731		328.639		306.829		571.440		546.197		173.634	
56. SO2 emissions	8137.068		6945.868		6509.985		12259.915		11752.624		8254.522	
57. CO emissions	767.921		265.448		247.912		461.943		441.665		178.671	
58. NOx emissions	1208.690		467.266		438.271		826.413		792.670		1058.604	
59. Regional employment	14886.242		38314.004		35479.642		69904.075		65888.358		59477.541	

* Industry outputs and final demands are measured in millions of dollars; energy consumption of each fuel type is measured in 10^{15} btu and pollution emissions are measured in tons.

6. Planning analysis for energy park development

6.1. INTRODUCTION

In the preceding chapters the energy park concept has been outlined in some detail (chapter 4). Moreover, a number of future scenarios for energy park development in the Montour region were composed in chapter 5. Based upon the information of these two chapters, we now consider application of the policy programming framework developed in chapter 3 to analyze the alternative options for energy park development in the Montour region.

We recall from Part One that the policy programming procedure was formulated to allow one to couple the interests of a number of policy makers to a model of system process functions. In chapter 4, we found that this set of process functions and the number of policy-making interests (collectively viewed as a policy-making system) would be remarkably large for the energy park planning problem. However, we also found that the issues of *practicality* are currently of most concern in assessing overall feasibility of energy parks at candidate sites.

Many of these issues of practicality center around regional economic, environmental, energy, and employment impacts. Hence, we found it convenient to restrict the overall planning problem to consideration of these impacts. By restricting the problem in this manner we note that the system we consider becomes a special case of the generalized *policy-making system* which was called an *energy-environment system*. In *energy-environment systems* the policy makers, system variables and system process functions are those variables and functions closely related to interindustry activity. The interindustry activity generated by energy park development in a region would account for many of the regional impacts associated with that development.

In chapter 3 a method of operationalizing the hierarchy of policy makers and their objectives in an energy-environment system was formalized – *the modified hierarchical model.* In addition, a model of the process functions in such a system was formulated – *the generalized input-output model.* The problem was then to determine a *composite scenario* for the region that is both collectively preferred by the policy-making interests being considered and *consistent* with the model of system process functions. The policy programming framework was developed for this purpose.

In chapter 5 the input-output model for the planning region (Montour) was developed to fill the requirements of the generalized input-output model of chapter 3. Consequently, the set of *future scenarios* corresponding to alternative energy park development plans for the Montour region were composed to be *consistent* with the input-output model.

In the following sections we apply the policy programming procedure using the set of *future scenarios* and the input-output model of the energy-environment system process functions.

6.2. APPLICATION OF POLICY PROGRAMMING

In order to apply the policy programming procedure, we must first determine the policy-making interests that we will consider. We recall from chapter 5 that the complete set of policy makers that are considered are those identified in Denton et al. [5] (see table 4.1). In order to reduce the scale of computation this complete set was aggregated into a number of *policy-maker groups* as follows:

1. Electric utility
2. Residential consumer
3. Industrial consumer
4. Environmentalist
5. Environmental regulator
6. Executive government
7. Legislative government
8. Oil and gas supplier
9. Coal supplier
10. Nuclear supplier.

In this aggregation the attempt was made to combine those policy makers with similar objectives into corresponding groups. In the policy programming framework we are also required to identify relevant *clusters* of system variables over which the policy-maker groups have varying degrees of influence. For reduction of computational requirements the total set of system variables was aggregated into *clusters* as follows:

1. Industry outputs
2. Wholesale and retail trade
3. Finance and services
4. Utility and government expenditures
5. Energy consumption
6. Environmental pollution
7. Employment.

We recall that the total set of system variables includes all the variables of a *scenario*: energy consumption levels, pollution emission levels, employment levels and industry output levels. We recall also from chapter 3 that the procedure for developing a *composite scenario* (the modified hierarchical model) is first to construct pairwise comparison matrices (to facilitate using the eigenvalue prioritization model) of the objectives of each policy-maker group. Then another set of these matrices are constructed to compare the relative contribution that each *future scenario* would make toward fulfilling the objectives of each policy-maker group. Identification of the policy-maker objectives was also taken from the Regional Assessment Study (Denton et al. [6]). These are summarized in table 6.1. The result of straight-forward hierarchical application of the eigenvalue model is a matrix, a column of which is the relative preference (a weighting vector) for the future scenarios by a policy-maker group. In the present case, the matrix is of order 6 by 10 (six future scenarios and ten policy-maker groups). This matrix (Q) is given in table 6.2.

If we knew the relative dominance of each policy-maker group with respect to 'overall influence' over the energy-environment system, then the composite scenario could easily be constructed by multiplying (Q) by the vector of relative weights corresponding to this measure of overall influence. The result would be a weighting vector of 'collective preference' for the future scenarios for all policy-maker groups. Multiplication of the matrix of future scenarios by this vector would yield a *composite scenario*. We recall that such a scenario would be consistent since it satisfies *Theorem 2* of chapter 3. However, as we also found in chapter 3, a much more realistic approach to constructing the composite scenario is to look at the relative influence various policy-maker groups have over the *clusters* of system variables. In the present case we wish to determine the relative influence of the ten actor groups over the seven clusters of system variables.

Derivation of these weights of relative influence is determined by constructing pairwise comparison matrices that weigh the relative influence of policy-maker groups for each cluster. Solution of the eigenvalue problem for each of these matrices yields the normalized weighting vectors of

TABLE 6.1.   Objectives of policy-maker groups.

1. Electric utility:
    1. Expand sales of electricity.
    2. Minimize cost and maintain 'fair' return.
    3. Encourage less stringent pollution emission regulation.
    4. Encourage less stringent safety regulation.
    5. Improve reliability.

2. Residential consumer:
    1. Encourage lower fuel prices.
    2. Encourage lower electricity prices.
    3. Lower consumer prices.
    4. Encourage 'substantial' supply of goods and services.
    5. Maintain environmental quality.

3. Industrial consumer:
    1. Encourage lower fuel prices.
    2. Encourage lower electricity prices.
    3. Encourage's 'substantial' supply of production materials.
    4. Maximize profit.
    5. Lower degree of government regulation.

4. Environmentalist:
    1. Reduce pollution emission levels.
    2. Maximize health and safety standards.

5. Environmental regulator:
    1. Maintain environmental quality.
    2. Enhance environmental quality.
    3. Conserve natural resources.

6. Executive government:
    1. Protect consumer interests.
    2. Protect regional industry interests.
    3. Maintain health, safety, and regional welfare.

7. Legislative government (same as executive government).

8. Oil and gas supplier:
    1. Maximize profit.
    2. Encourage less stringent gas and oil regulation.
    3. Discourage strict environmental regulation.

9. Coal supplier:
    1. Maximize profit.
    2. Encourage less strict environmental regulation.
    3. Develop steady long-term markets for coal.

10. Nuclear suppliers:
    1. Maximize profit.
    2. Encourage less stringent environmental regulation.
    3. Develop steady uranium markets.

TABLE 6.2. Policy-maker future scenario weights.

Future scenarios	Electric utility	Residential consumer	Industrial consumer	Environmentalist	Environmental regulator	Executive government	Legislative government	Oil and gas supplier	Coal supplier	Nuclear supplier
1	0.08	0.13	0.08	0.11	0.16	0.05	0.04	0.14	0.15	0.12
2	0.31	0.28	0.19	0.25	0.27	0.32	0.27	0.06	0.07	0.41
3	0.03	0.15	0.06	0.37	0.29	0.04	0.07	0.03	0.03	0.08
4	0.34	0.27	0.17	0.20	0.21	0.25	0.21	0.05	0.06	0.33
5	0.17	0.08	0.25	0.02	0.03	0.20	0.25	0.11	0.60	0.03
6	0.07	0.09	0.25	0.05	0.05	0.13	0.15	0.62	0.09	0.04
	1.00	1.00	1.00	1.00	1.00	1.00	1.00	1.00	1.00	1.00

Note: Each actor compares his objectives (an o_i by o_i matrix) which, upon solving the eigenvalue problem yields an o_i by 1 vector (the eigenvector). He then compares the relative desirability of each future scenario for each of his objectives (a 6 by 6 matrix) which, upon once again solving the eigenvalue problem, yields a set of 6 by 1 vectors. The matrix of relative weights is constructed by placing the scenario weights as columns (an o_i by 6 matrix). Upon multiplying the matrix of weights by the objectives an overall weighting vector of the scenarios for each policy maker results. These vectors are the columns of the matrix D given in the table.

influence (eigenvectors). The results of solving the eigenvalue problems for each of these matrices are given in table 6.3.

We recall from chapter 2 that overall consistency of the judgements in pairwise comparison matrices can be measured by the value $\alpha = (\lambda_{max} - n)/(n - 1)$, where λ_{max} is the dominant eigenvalue of the matrix and n is the dimension of that matrix. The closer α is to zero the more consistent the matrix of judgements is presumed to be. A 'map' of the pairwise comparison matrices required in the current application of the modified hierarchical model is shown in fig. 6.1. The indices of consistency for these matrices are given in table 6.4. Saaty [49] describes acceptable values of α to be those that are significantly closer to zero than those of a randomly filled matrix[1] of pairwise comparisons of the same order. The mean values of for twenty random (in the same sense) matrices of order corresponding to each of the pairwise matrices used in the hierarchical model are also given in table 6.4. All α's of the pairwise comparison matrices are significantly closer to zero than the mean of the corresponding random matrices.

We next construct the matrix of *preference scenarios* by (as we recall from chapter 3):

$$\mathbf{F}_i = \mathcal{A}\,\mathbf{Q}_i \qquad\qquad\qquad i = 1, 2, \ldots, n$$

where $\mathcal{A}$ is the matrix of *future scenarios* (each column is a future scenario) and $\mathbf{F}_i$ is the *preference scenario* for policy maker group i. F (the matrix of *preference scenarios*) is simply the matrix formed by placing the $\mathbf{F}_i$'s as columns (see table 6.5).

Finally, we can construct the *composite scenario* by multiplying each row of the preference scenario matrix by the appropriate row of G; i.e. each row of F corresponds to a particular system variable which is a member of one of the aggregated variable clusters. To determine the composite value of that variable, one takes the sum of the preferred values of the variable multiplied by the relative influence each group has over the cluster of which the variable is a member. The collection of all such composite values for system variable forms a *composite scenario*. This scenario, we recall from chapter 3, is the set of planning targets collectively preferred by all policy-making groups based upon the objectives and influence of the policy-making groups alone. The composite scenario for the present case is given in table 6.6.

---

1. These 'random' matrices are not really random since they are still required to satisfy $a_{ji} = 1/a_{ij}$.

TABLE 6.3. Policy-maker influence over system variable clusters.

Variable cluster	Electric utility	Residential consumer	Industrial consumer	Environmentalist	Environmental regulator	Executive government	Legislative government	Oil and gas suppliers	Coal supplier	Nuclear supplier
1	0.06	0.02	0.27	0.02	0.07	0.20	0.15	0.13	0.06	0.02
2	0.15	0.34	0.17	0.01	0.03	0.11	0.09	0.05	0.03	0.02
3	0.04	0.02	0.06	0.07	0.26	0.22	0.13	0.06	0.10	0.04
4	0.04	0.02	0.06	0.07	0.26	0.22	0.13	0.06	0.10	0.04
5	0.16	0.02	0.12	0.02	0.04	0.10	0.06	0.23	0.19	0.05
6	0.06	0.05	0.20	0.01	0.03	0.26	0.13	0.17	0.06	0.02
7	0.15	0.34	0.17	0.01	0.03	0.11	0.09	0.05	0.03	0.02

FIG. 6.1. Modified hierarchical model pairwise comparison matrix map.

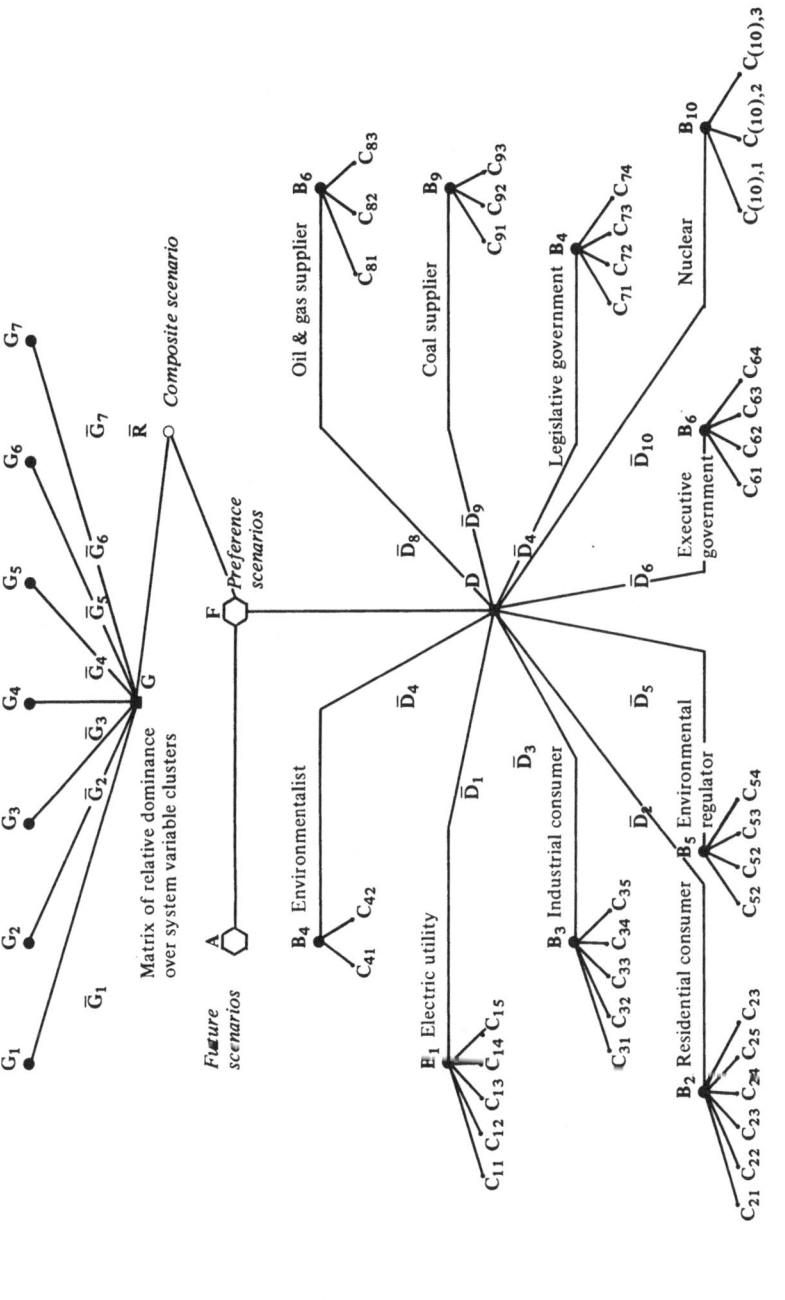

O_e — Comparison matrix of policy maker influence over *cluster* e.
O_{ij} — Comparison matrix of *future scenario* contribution to objective j of policy maker i.
B_i — Comparison matrix of objectives for policy maker i.
Note: For a better discussion of the notation in this table see Table 3.1.

● — Matrix of pairwise comparisons
■ — Matrix of concatenated normalized eigenvectors
○ — Composite scenario (not consistent)
⬡ — Matrix of consistent scenarios

TABLE 6.4.   Pairwise comparison matrix judgement consistency.

Policy maker	Pairwise comparison matrix	$(\lambda_{max}-n)/n$	α_{rand}	α^*
1. Electric utility (n = 5)	B_1	.30	.32	.12
	C_{11}	.06	.37	
	C_{12}	.05	.37	
	C_{13}	.26	.37	
	C_{14}	.08	.37	
	C_{15}	.07	.37	
2. Residential consumer (n = 5)	B_2	.11	.32	
	C_{21}	.13	.37	
	C_{22}	.09	.37	
	C_{23}	.29	.37	
	C_{24}	.17	.37	
	C_{25}	.10	.37	
3. Industrial consumer (n = 4)	B_3	.20	.32	
	C_{31}	.11	.37	
	C_{32}	.11	.37	
	C_{33}	.13	.37	
	C_{34}	.09	.37	
4. Environmentalist (n = 2)	B_4	.00	.00**	
	C_{41}	.05	.37	
	C_{42}	.08	.37	
5. Environmental regulator (n = 4)	B_5	.12	.28	
	C_{51}	.29	.37	.15
	C_{52}	.06	.37	
	C_{53}	.13	.37	
	C_{54}	.32	.37	
6. Executive government (n = 4)	B_6	.09	.28	
	C_{61}	.07	.37	
	C_{62}	.09	.37	
	C_{63}	.05	.37	
	C_{64}	.07	.37	
7. Legislative government (n = 4)	B_7	.11	.28	
	C_{71}	.61	.37	
	C_{72}	.09	.37	.13
	C_{73}	.10	.37	
	C_{74}	.14	.37	
8. Oil and Gas supplier (n = 3)	B_8	.06	.13	
	C_{81}	.09	.37	
	C_{82}	.10	.37	
	C_{83}	.13	.37	
9. Coal supplier (n = 3)	B_9	.01	.13	
	C_{91}	.11	.37	
	C_{92}	.15	.37	
	C_{93}	.21	.37	
10. Nuclear supplier (n = 3)	B_{10}	.05	.13	
	$C_{10,1}$	.07	.37	
	$C_{10,2}$	.13	.37	
	$C_{10,3}$	.05	.37	

*   In cases where the index of consistency $(\lambda_{max} - n)/n$ is not acceptable, a revised matrix of pairwise comparisons was used.
**   A second order matrix is always consistent.

As was discussed in previous chapters and illustrated in the recurring example of those chapters, this composite scenario is not necessarily consistent. That is, when we account for the relationships among variables in the energy-environment system which are captured in the generalized input-output model of the system process functions, we find that the requirements for a *consistent scenario* (*Definition 2* of chapter 3) are not satisfied. However, as we found in the development of policy programming procedure of chapter 3, we can systematically alter the planning targets given in the *composite scenario* until a *consistent composite scenario* results. In the next section, before we perform this alteration, we review the individual and collective planning targets that have been developed thus far, i.e. the *preference* (table 6.5) and *composite* (table 6.6) scenarios, respectively.

6.3. OBSERVATIONS ON SCENARIOS

In the process of applying the modified hierarchical model to obtain the planning targets (*composite scenario*) four classes of scenarios have been developed: (1) *future*, (2) *reference*, (3) *composite* and (4) *planning* (consistent composite). The *future scenarios* were constructed to portray *consistent* projections of regional economic, energy, environmental, and employment trends corresponding to alternative energy park development plans.

The *preference scenarios*, one for each policy-maker group, are linear combinations of the *future scenarios*. The weights for the linear combinations are provided by the degree to which a *future scenario* contributes to a policy maker's objectives. The collection of *preference scenarios* for the energy park planning problem presents some interesting but expected contrasts. For instance, by comparing the *preference scenario* of the *residential consumer* to the set of *future scenarios* we see that the residential consumers are willing to settle for a lower level of employment than might be possible by favoring some other *future scenarios*, perhaps due to the fact that regional environmental quality would be compromised by the higher employment scenarios (e.g. the 20GW$_e$ energy park plans).

We can observe from table 6.5, or perhaps more easily from table 6.2, what the *preference scenarios* of each policy-maker group reflects. The *electric utility* favors, as one might expect, either of the nuclear scenarios (2 and 4) since they provide the most economical energy park development options. The *industrial consumer* favors the large hybrid park scenarios (5 and 6) since these scenarios offer the most potential for

TABLE 6.5. Preference scenarios*

Sector classification	Electric utility	Resid. consumer	Indust. consumer	Environ- mentalist
1. Mining	5.57	5.05	5.60	4.52
2. Food processing	50.90	43.94	42.16	41.62
3. Textiles	12.70	12.49	11.07	11.41
4. Matr. processing	26.94	27.65	28.20	24.30
5. Prtg. publish	5.18	4.57	4.90	4.19
6. Fabr., assem.	117.95	116.24	177.30	94.72
7. Chemicals	8.03	7.80	7.08	7.13
8. Sawmills	1.63	1.41	1.43	1.31
9. Pulpwood	0.24	0.23	0.20	0.21
10. Agr. feeds, fert.	2.67	2.35	2.27	2.21
11. Misc. industry	352.80	310.86	370.31	277.49
12. Agriculture	21.88	19.04	18.37	17.95
13. Education	3.13	2.87	2.79	2.65
14. Foodstores	62.42	55.18	58.70	50.61
15. Gas stations	30.50	27.02	29.33	24.61
16. Auto dealers	62.17	55.25	60.20	50.22
17. Clothing	29.44	25.69	25.98	23.96
18. Bars, restaurants	25.97	22.66	22.91	21.16
19. Jwlry., drgs.	13.33	11.67	11.97	10.82
20. Furn. and appl.	21.92	19.36	20.53	17.77
21. Hrdware, bld. mat.	28.89	25.63	28.48	23.23
22. Depart., variety	37.40	32.92	34.49	30.35
23. Farm equipment	6.36	5.61	5.76	5.18
24. All other retail	8.41	7.46	7.60	6.90
25. Hotels, motels	15.61	13.50	13.22	12.75
26. Enter., recreation	2.07	1.85	1.96	1.70
27. Finance	24.12	20.99	21.07	19.62
28. Rl. estate insur.	52.47	45.74	48.98	42.13
29. Laundry	0.57	0.54	0.56	0.48
30. Prof. pers. svcs.	137.06	119.06	130.53	106.10
31. Nonprof.pers.svcs.	12.40	10.95	11.45	10.09
32. Transportation	68.32	60.08	67.31	54.07
33. Construction	42.58	37.36	40.46	33.84
34. Wholesale	65.20	57.92	62.23	52.89
35. Electric, gas	10.95	9.79	10.42	8.96
36. Phone	21.59	18.91	20.82	17.17
37. Water, tv, radio	5.05	4.46	4.66	4.11
38. Nonprofit	37.56	33.08	34.55	30.49
39. Local gvt.	18.05	15.58	15.16	14.72
40. Public schools	12.86	11.67	12.81	10.52
41. County govt.	2.13	1.94	2.18	1.74
42. State govt.	3.80	3.42	3.44	3.15
43. Federal govt.	5.58	4.81	4.61	4.56
44. Labor	186.04	170.94	203.32	110.93
45. Rent	77.93	66.80	63.65	63.52
46. Transfer	9.25	8.98	10.28	7.87
47. Proprietary	102.78	90.37	100.29	81.65
48. Overflow	7.83	7.53	10.40	6.36
49. Coal	9.17	8.18	9.16	7.39
50. Crude oil and gas	34.58	30.61	35.30	27.52
51. Refined petroleum	16.18	14.32	15.90	12.95
52. Electric utilities	1.71	1.54	1.80	1.38
53. Natural gas utilities	12.97	11.62	13.41	10.42
54. Part. emissions	9478.79	9419.68	9537.60	8385.29
55. SO 2 emissions	489.47	476.19	435.65	434.25
56. CO emissions	9745.49	8979.32	9451.58	8144.11
57. NO X emissions	398.01	387.96	361.57	352.54
58. Regional employ-ment	744.39	733.11	816.10	642.79
	53115.85	47487.75	53770.57	42639.18

* Industry outputs and final demands are measured in millions of dollars; energy consumption is measured in 10^{15} btu and pollution emissions in tons.

TABLE 6.5. Preference scenarios (continued)

Environ. regulator	Government executive	legis.	Oil & gas	Energy Supplier Coal	Nuclear
4.56	5.54	5.59	5.46	5.93	5.03
40.51	48.12	47.93	21.30	50.60	46.23
12.49	11.34	11.06	8.64	14.53	12.75
27.15	25.51	25.26	33.12	32.22	26.99
4.12	5.10	5.13	4.02	5.32	4.65
96.34	139.56	146.08	293.00	133.90	101.78
7.72	7.25	7.10	5.59	9.07	7.96
1.28	1.57	1.57	0.91	1.64	1.46
0.22	0.21	0.21	0.13	0.26	0.23
2.19	2.52	2.51	1.35	2.72	2.45
269.83	363.29	370.45	377.60	367.37	307.09
17.60	20.70	20.61	10.06	21.95	19.95
2.72	2.94	2.92	2.07	3.33	2.95
49.87	61.31	61.65	47.86	63.82	56.23
24.27	30.19	30.42	25.34	31.32	27.38
49.65	61.61	62.10	53.00	64.14	55.84
23.44	28.37	28.39	17.24	29.65	26.62
20.70	25.03	25.05	15.19	26.12	23.49
10.61	12.91	12.93	8.51	13.44	12.04
17.50	21.50	21.61	16.55	22.39	19.75
22.88	28.88	29.18	25.99	29.84	25.78
29.84	36.51	36.66	26.55	38.09	33.72
5.12	6.15	6.16	4.22	6.50	5.76
6.85	8.11	8.11	5.60	8.63	7.67
12.39	14.89	14.86	7.38	15.53	14.13
1.69	2.03	2.04	1.65	2.13	1.88
19.17	23.17	23.18	13.44	24.26	21.80
41.05	51.63	52.09	37.89	54.00	46.61
0.50	0.55	0.55	0.56	0.62	0.53
103.93	135.91	136.17	114.92	131.35	122.40
9.95	12.09	12.14	8.87	12.68	11.21
53.12	68.30	69.11	60.90	70.84	60.39
33.31	41.98	42.26	33.95	43.49	37.99
52.31	64.26	64.68	52.91	67.15	58.76
8.91	10.75	10.81	8.82	11.37	9.93
16.80	21.47	21.68	17.89	22.09	19.15
4.06	4.92	4.94	3.61	5.17	4.57
29.99	36.63	36.75	26.52	38.10	33.93
14.31	17.16	17.12	8.17	17.95	16.35
10.58	12.77	12.87	12.26	13.54	11.69
1.75	2.14	2.16	2.21	2.25	1.93
3.17	3.64	3.63	2.61	3.96	3.50
4.44	5.27	5.25	2.31	5.55	5.07
151.77	191.64	194.93	230.54	199.47	166.64
61.49	73.54	73.24	29.52	76.90	70.57
8.30	9.35	9.44	12.50	10.41	8.68
80.05	102.61	103.55	89.75	104.32	91.42
6.45	8.75	9.05	15.33	8.68	6.90
7.30	9.20	9.31	8.62	9.57	8.18
26.95	35.12	35.66	34.35	35.98	30.47
12.76	16.15	16.31	14.40	16.71	14.42
1.37	1.75	1.77	1.89	1.80	1.52
10.29	13.18	13.38	13.52	13.58	11.52
9141.27	8987.50	8915.60	10202.13	10952.68	9312.56
470.39	444.02	435.30	353.04	553.64	484.81
8320.94	9492.13	9513.17	8673.91	10425.22	9031.34
381.76	363.98	357.65	308.81	451.00	393.22
688.68	740.44	744.36	985.10	851.99	710.17
42212.43	53512.33	54124.35	52530.63	54953.39	47463.74

TABLE 6.6. Composite scenario*

Sector classification	X (industry outputs)	State variables	X* (system variables)
1. Mining	5.47	Coal	8.838
2. Food processing	42.60	Crude oil and gas	34.063
3. Textiles	11.29	Refined petroleum	15.509
4. Matr. processing	27.84	Electric utilities	1.722
5. Prtg. publish	4.83	Natural gas utilities	12.883
6. Fabr., assem.	163.73	Part emissions	7211.277
7. Chemicals	7.18	HC emissions	436.206
8. Sawmills	1.42	SO_2 emissions	8772.056
9. Pulpwood	0.20	CO emissions	350.889
10. Agr. feeds, fert.	2.29	NO_x emissions	644.582
11. Misc. industry	357.51	Regional employ-	
12. Agriculture	18.53	ment	51322.084
13. Education	2.88		
14. Foodstores	57.87		
15. Gas stations	28.54		
16. Auto dealers	58.40		
17. Clothing	26.51		
18. Bars, restaurants	23.39		
19. Jwlry., drgs.	12.10		
20. Furn. and appl.	20.28		
21. Hrdware, bld. mat.	27.32		
22. Depart., variety	34.36		
23. Farm equipment	5.80		
24. All other retail	7.69		
25. Hotels, motels	13.80		
26. Enter., recreation	1.93		
27. Finance	21.20		
28. Rl. estate insur.	47.20		
29. Laundry	0.55		
30. Prof. pers. svcs.	122.63		
31. Nonprof. pers. svcs.	11.17		
32. Transportation	65.73		
33. Construction	39.79		
34. Wholesale	61.29		
35. Electric, gas	10.03		
36. Phone	19.63		
37. Water, tv, radio	4.55		
38. Nonprofit	33.74		
39. Local govt.	15.62		
40. Public schools	12.03		
41. County govt.	2.02		
42. State govt.	3.43		
43. Federal govt.	4.80		
44. Labor	185.80		
45. Rent	66.62		
46. Transfer	9.29		
47. Proprietary	93.91		
48. Overflow	8.38		

* Industry outputs are measured in millions of dollars; energy consumption is measured in 10^{15} btu and pollution emissions in tons.

providing markets for regionally produced goods and services. The environmental policy makers (*environmentalist* and *regulator*) both favor the nuclear and small hybrid parks since the large hybrid park alternatives (5 and 6) present unacceptable SO_2 emission levels. The energy suppliers (*oil and gas*, *coal* and *nuclear*) all favor the scenarios that promote their respective products. The government policy makers (*executive* and *legislative*) favor the nuclear and large hybrid fossil alternatives apparently since these scenarios offer the highest potential economic and employment benefits to the region.

The *composite scenario* is finally derived as a combination of the *preference scenarios* weighted by the influence that the policy makers have over the system variable *clusters*. We can very easily verify that the *composite scenario* (table 6.6) is not consistent. Table 6.7 shows the values of the state variables X^* from the *composite scenario*. Alongside these values in the table are the corresponding values of those variables computed by taking the product DX. We shall call this collection of state variables $X^{*\prime}$. We recall from chapter 3 that the *composite scenario* is consistent only if $X^* = X^{*\prime}$:

For $\mathcal{X} = (X, X^*)$, a scenario

$\quad$ $X^* = D\,X$ for $\mathcal{X}$ to be *consistent*

but $\quad X^{*\prime} = D\,X$

so $\quad X^{*\prime} \neq X^*$

An interesting point concerning the comparison of X^* and $X^{*\prime}$ for the specific case in table 6.7 is that the values of energy and employment

TABLE 6.7. Composite scenario state variables.

	X^*	$X^{*\prime}$	
Coal	8.838	8.950	
Crude oil and gas	34.063	34.300	
Refined petroleum	15.509	15.520	10^6 Btu
Electric utilities	1.722	1.760	
Natural gas utilities	12.883	13.070	
Part emissions	7211.277	9468.650	
HC emissions	436.206	438.710	
SO_2 emissions	8772.056	9330.620	tons
CO emissions	350.889	363.130	
NO_x emissions	644.582	800.220	
Regional employment	51322.084	50898.870	man-years

variables for each are very close (nearly consistent). However, the environmental variables are quite far off. This indicates that the policy-making interests in the region prefer both high levels of industrial production and low levels of pollution emission. With current technology (reflected in the input-output model of system process functions), this cannot be permitted.

Hence, we must invoke the goal programming algorithm in order to find the most suitable *consistent composite scenario* in terms of the priorities and influence of policy makers in the regional energy-environment system.

6.4. THE CONSISTENT COMPOSITE SCENARIO

We recall from chapter 3 that solution of the following goal programming problem will yield the *planning scenario* (consistent composite):

$$\min Z = \mathbf{P_+ d^-} + \mathbf{P_- d^-} \qquad (6.4.1)$$

$$\text{subject to} \quad \mathbf{BX} + \mathbf{d_1^-} - \mathbf{d_1^+} = \mathbf{\bar{R}}$$

$$\mathbf{BX} + \mathbf{d_{i+1}^3} - \mathbf{d_{i+1}^+} = \mathbf{F}_i \qquad\qquad i = 1, 2, \ldots, n.$$

If we were to attempt to construct the entire simplex tableau for this problem (for the energy park problem) this would require a 656 by 1347 matrix. However, we can dismiss many rows and columns by neglecting low priority objectives of individual policy makers.

We can also take another shortcut by posing high priority objectives of the most influencial individual policy makers in terms of the deviational variables of the collective objectives (equation 6.4.1). In fact, if we assume that all individual objectives can be posed in terms of the deviational variables of collective objectives, we can reduce the simplex tableau to a more reasonable 66 by 167 matrix. The goal programming objective function for the reduced tableau is shown in table 6.8.

The resulting *consistent composite scenario* for this problem is shown in table 6.9.

TABLE 6.8. Goal programming objective function.

$$Z = P_1 \left[62(d_1^- + d_2^- + d_3^- + d_4^- + d_5^- + d_6^- + d_7^- + d_8^- + d_9^- + d_{10}^- + d_{11}^- + d_{12}^-) + 20(d_1^+ + d_2^+ + d_3^+ + d_4^+ + \right.$$
$$\left. + 20(d_1^+ + d_2^+ + d_4^+ + d_5^+ + d_6^+ + d_7^+ + d_8^+ + d_9^+ + d_{10}^+ + d_{11}^+ + d_{12}^+) \right]$$
$$+ P_2 \left[d_{13}^- + d_{14}^- + d_{15}^- + d_{16}^- + d_{17}^- + d_{19}^- + d_{20}^- + d_{21}^- + d_{22}^- + d_{23}^- + d_{24}^- + d_{25}^- + d_{26}^- \right]$$
$$+ P_3 \left[22(d_{27}^- + d_{28}^-) + 25(d_{30}^- + d_{31}^-) \right]$$
$$+ P_4 \left[22(d_{25}^- + d_{36}^- + d_{37}^- + d_{38}^-) + 35(d_{39}^- + d_{40}^- + d_{41}^- + d_{42}^- + d_{43}^-) \right]$$
$$+ P_5 \left[23(d_{49}^-) + 18(d_{50}^- + d_{51}^- + d_{53}^-) + 16(d_{52}^-) \right]$$
$$+ P_6 \left[26(d_{54}^2 + d_{55}^2 + d_{56}^{\bar{2}} + d_{57}^2 + d_{58}^2) + 37(d_{54}^{d3} + d_{55}^3 + d_{56}^3 + d_{57}^3 + d_{58}^{\pounds}) \right]$$
$$+ P_7 (d_{59}^-)$$

TABLE 6.9.   Consistent composite planning scenario*

Sector classification	X	State variables	X*
1. Mining	5.484	Coal	8.849
2. Food processing	41.098	Crude oil and gas	34.060
3. Textiles	12.332	Refined pretroleum	15.518
4. Matr. processing	27.025	Electric utilities	1.742
5. Prtg. publish	6.378	Natural gas utilities	13.015
6. Fabr., assem.	162.909	Part emissions	9448.711
7. Chemicals	6.105	HC emissions	441.497
8. Sawmills	0.884	SO_2 emissions	9275.908
9. Pulpwood	0.665	CO emissions	364.896
10. Agr. feeds, fert.	0.830	NO_x emissions	793.967
11. Misc. industry	358.985	Regional employ-	
12. Agriculture	17.959	ment	51576.172
13. Education	2.641		
14. Foodstores	58.148		
15. Gas stations	28.493		
16. Auto dealers	57.523		
17. Clothing	25.294		
18. Bars, restaurants	21.697		
19. Jwlry., drgs.	12.788		
20. Furn. and appl.	18.421		
21. Hrdware, bld. mat.	27.909		
22. Depart., variety	32.888		
23. Farm equipment	3.832		
24. All other retail	9.550		
25. Hotels, motels	15.383		
26. Enter., recreation	1.871		
27. Finance	22.982		
28. Rl. estate insur.	46.478		
29. Laundry	0.872		
30. Prof. pers. svcs.	123.439		
31. Nonprof. pers. svcs.	12.653		
32. Transportation	64.824		
33. Construction	40.774		
34. Wholesale	61.056		
35. Electric, gas	8.307		
36. Phone	21.104		
37. Water, tv, radio	3.412		
38. Nonprofit	34.168		
39. Local govt.	16.957		
40. Public schools	12.875		
41. County govt.	3.548		
42. State govt.	3.344		
43. Federal govt.	3.403		
44. Labor	185.745		
45. Rent	68.240		
46. Transfer	7.583		
47. Proprietary	92.334		
48. Overflow	7.625		

* Industry outputs are measured in millions of dollars; energy consumption is measured in 10^{15} btu and pollution emissions are measured in tons.

7. Conclusions and recommendations for further research

7.1. ENERGY PARK EVALUATION

The dramatic difference between the *composite* and *consistent composite scenarios* for the Montour region implies a very important conclusion concerning energy park development in that region. The great differences between many values of variables in the two scenarios indicates, as mentioned earlier, that the scenario collectively preferred by policy makers in the region (the composite scenario) is not feasible in terms of the model of system process functions. In particular, this scenario reflects high levels of industrial production and employment as well as low levels of air pollution emission associated with that production. The technical relationships between the economic production and pollution emission will not permit such a scenario.

The policy programming procedure allows us to relax the planning targets given by the composite scenario (in reverse order of priority) until a consistent scenario is achieved. However, this new scenario displays significant compromises in the originally preferred levels of pollution emissions. For some industries, the objectives of improving the economic health in the region promoted increased production and employment in those industries. However, for other industries, the reverse was true, i.e. environmental objectives constrained production for those industries.

Hence, in order to implement a plan that would bring about the *consistent composite scenario* in the region, the role of regional planning agencies and government in this implementation would have to be significantly larger than that suggested by many energy park development studies (see e.g. General Electric [68], Pennsylvania Energy Park Development Group [85], NUREG-0001 [110]). The plan for bringing about the consistent composite scenario would be much broader than siting, constructing and operating the various park facilities. Controls would have to be implemented to regulate the source of materials for those projects. The electric utility, or group of utilities, alone could not effectively bring about the *planning scenario*. Some regional planning agency involvement would be required to implement measures to restrict the output of some industries while stimulating the outputs of others.

This goes far beyond the 'compacts' for interstate cooperation in energy park development suggested by the General Electric study mentioned in chapter 4. Indeed, the implication is that the government or regional planning agency's role in planning for energy park development be elevated to equal or greater magnitude than that of the electric utility. While not all of the issues concerning energy park development have been considered incoming to this conclusion, it is clear that in some regions the impact of a large energy park would be so significant as to transform completely the region's socioeconomic profile. An electric utility alone could not effectively control this transformation in the best interests of the region.

Table 7.1 shows the state variables of the (1) *composite scenario* X^*, (2) the planning scenario X^*_{cons}, and (3) the consistent values of state variables corresponding to the total outputs of the composite scenario, i.e. the product $DX = X^{*'}$.

From the table we can easily see where the collective priorities of the region rest. The *composite scenario* reflects preference for somewhat lower values of energy consumption and higher levels of employment than those that are consistent with the economic outputs of that scenario, i.e. $X^{*'}$.

In the modification of the composite scenario to consistency, we notice (in the X^*_{cons} column of table 7.2) that high priority was apparently placed on achieving the economic and energy consumption target levels while a

TABLE 7.1. Planning scenario consistency.

	X^*	X^*_{cons}	$X^{*'}$	
Coal	8,838	8.849	8.950	
Crude oil and gas	34.063	34.060	34.300	
Refined petroleum	15.509	15.518	15.520	10^{15} Btu
Electric utilities	1.722	1.742	1.760	
Natural gas utilities	12.883	13.015	13.070	
Part emissions	7211.277	9448.711	9469.650	
HC emissions	436.206	441.497	438.710	
SO_2 emissions	8772.056	9275.908	9330.620	tons
CO emissions	350.889	364.896	363.130	
NO_x emissions	644.582	793.967	800.220	
Regional employment	51322.084	51576.172	50898.870	man-years

TABLE 7.2. Composite scenario final demands.

Sector classification	Composite scenario $Y(10^6\$)$	Planning scenario $Y(10^6\$)$
1. Mining	0.206	0.190
2. Food Processing	30.413	28.960
3. Textiles	4.237	5.159
4. Matr. processing	7.985	7.823
5. Prtg. publish	0.637	2.292
6. Fabr., assem.	116.719	104.618
7. Chemicals	0.892	0.198
8. Sawmills	0.679	0.147
9. Pulpwood	0.173	0.634
10. Agr. feeds, fert.	0.171	1.229
11. Misc. industry	354.872	356.352
12. Agriculture	4.945	4.856
13. Education	1.544	1.311
14. Foodstores	9.041	9.857
15. Gas stations	1.965	2.226
16. Auto dealers	2.631	2.204
17. Clothing	13.038	11.910
18. Bars, restaurants	13.414	11.868
19. Jwlry., drgs.	4.947	5.660
20. Furn. and appl.	4.344	2.802
21. Hrdware, bld. mat.	1.773	2.376
22. Depart., variety	9.392	8.087
23. Farm equipment	0.659	1.293
24. All other retail	0.326	2.330
25. Hotels, motels	11.789	13.346
26. Enter., recreation	0.894	0.839
27. Finance	10.772	12.700
28. Rl. estate insur.	19.479	19.016
29. Laundry	0.052	1.369
30. Prof. pers. svcs.	114.810	115.585
31. Nonprof. pers. svcs.	3.666	5.162
32. Transportation	23.965	23.004
33. Construction	16.129	16.620
34. Wholesale	9.913	10.095
35. Electric, gas	3.014	1.368
36. Phone	7.569	9.097
37. Water, tv, radio	0.548	0.458
38. Nonprofit	10.885	11.451
39. Local govt.	11.168	12.171
40. Public schools	1.023	1.969
41. County govt.	0.038	1.508
42. State govt.	0.593	0.591
43. Federal govt.	3.912	2.515
44. Labor	6.476	4.822
45. Rent	54.382	56.294
46. Transfer	0.100	1.969
47. Proprietary	3.129	4.352
48. Overflow	0.893	1.624

relatively much lower priority was placed on achieving the low pollution emission level targets reflected in X^*.

As a result the values of X^*_{cons} for energy consumption are fairly close to those of X^*. On the other hand, the values of pollution emission in X^*_{cons} are much higher than those indicated in X^*. This is, of course, due to the technical interdependence between pollution and industrial production (represented as equations in the model of system process functions) which required these increased emission levels since the achievement of economic targets received a higher priority than achievement of emissions targets.

Hence, inclusion of the model of system process functions and enforcing consistency of the *planning scenario* with this model dramatically brings out the tradeoffs that appear in problems such as planning for energy park development. In the present case, if regional planners wish to reap maximum economic benefits of energy park development in the region, they must be content with higher than preferred levels of pollution.

If we compare the list of final demands presented to the regional economy in the *planning scenario* (table 7.2) to the alternative *future scenarios* (table 5.15) we see that the former scenario corresponds fairly well with those of the small energy park options, i.e. the small NEC, the small hybrid park and the special hybrid park (*future scenarios* 2, 3 and 6, respectively).

In implementing a plan for developing one of these energy park options, a regional plan for controlling this development might include steps to control the materials and resident employment for the construction and operation of the park, thereby controlling the corresponding increased final demands presented to the regional economy. In this way, the planners might be able to 'fine tune' the development to come as close to the specifications of the *planning scenario*.

7.2. POLICY PROGRAMMING

The use of policy programming is, of course, not limited to energy planning problems. The policy programming framework is best suited to problems where the structure and flows of the system under study can be characterized by a suitable model of process functions such as the generalized input-output model presented in chapter 2. An important feature of the analysis by policy programming is that the objectives and influence of many policy-making interests can be captured. These considerations are treated as externalities or neglected entirely in many current energy planning models.

The viewing of planning problems in the context of the policy programming framework amounts to, perhaps, a fundamental shift in the analysis of such problems. Traditional approaches choose a planning option from a number of rigidly specified alternatives. Policy programming seeks to synthesize a collectively preferred planning option from a number of alternatives subject to the constraints that make the composite plan feasible (the model of system process functions).

7.3. RECOMMENDATIONS FOR FURTHER RESEARCH

The policy programming framework as it was implemented in this study deals only with the specific case of *energy-environment systems*, which we recall from chapter 2, forms a special case of the more general notion of *policy-making systems*.

A fundamental area of further research might be to find ways to operationalize the aspects of policy-making systems that were not captured in the model of system process functions for energy-environment systems. These aspects include such concepts as accounting for the problems associated with policy implementation or making policy programming dynamic.

With regard to energy parks, the results of this study indicate that much more research effort should be expended on the planning for energy park development in specific siting regions. The roles of government and regional planning agencies in such planning efforts must be given particular attention. Moreover, since the results of this study indicate that indirect socioeconomic impact on the region in which an energy park is sited could be quite extensive, the relationships between the energy park development project and regional industry should be examined in detail.

Part Three

Appendices

Appendices

Appendix A

Generalized input-output analysis

The extensions of input-output to investigating energy consumption, environmental pollution and employment generally relate the measures of these quantities (i.e. tons of coal, tons of SO_2 emitted, etc) linearly to total outputs (see Herendeen and Bullard [136], White [170], Toscas [164], Strout [163], and Reardon [159]). The results are often computed from impact intensity vectors or coefficients (e.g. tons/$). However, in the case of energy, special care must be taken since energy is often a pass-through commodity, as in the case of petroleum refining, where the industry receives crude oil but actually consumes very little of it. Hence, we require a modified approach to account for this secondary production of energy.

Energy intensity
A general approach to determining changes in energy consumption trends based on given changes in final demand has been by constructing a separate Total Energy Requirements Matrix:[1]

$$\Delta T = e\Delta Y$$

where e is a total energy requirements matrix and ΔT is the change in outputs of energy sectors required to support a change in final demand ΔY ($).

The procedure for finding the total dollar requirements was shown in chapter 2. We can also develop a scheme for determining a total energy requirements matrix.

In order to develop the total requirements matrix itself let us first develop a related concept, i.e. that of a table of direct energy flows or transactions. Determination of such a table is based on the availability of the following two bodies of data:

1. This is the approach adopted by Rearden [159], Herendeen [134], Just [146] and others.

1. The matrix of technical coefficients **A**.
2. A table of implied energy prices ($/Btu) that describes how much each producer paid to each of the energy sectors in the economy per unit of energy supplied (say Btu). Actually, of interest is a table of inverse prices (Btu/$) or the element-by-element inverse of energy prices. Let us call this table one of Implied Inverse Energy Prices **P**, which is of dimension $m \times n$ where m is the number of energy sectors in the economy and n is the total number of sectors included in the I/O table. It is important to note that calculation of **P** is generally performed by surveying to determine the flows of energy from energy-supplying sectors to energy-consuming sectors and then dividing each energy flow by its associated dollar transaction. This resulting set of 'prices' is implied in that it seldom reflects the actual price paid for energy.

Several published sources of Implied Inverse Energy Prices are available (e.g. Herendeen [134], which can be used to recover the energy flows – (originally used to derive energy prices) – in the following manner: if we let Z^e be the m energy rows of the transactions matrix (those rows of the energy supplying sectors of the economy), then the Z^e matrix, which is of dimension $m \times n$, describes the economic distribution of each energy sector's output. An element-by-element multiplication of X^e and **P** yields the direct flows of energy, **E**, which is also of dimension $m \times n$, from the energy sectors to all consuming sectors):

$$E_{kj}(\text{Btu}) = Z^e_{kj}(\$) \cdot P_{kj}(\text{Btu}/\$) \ k = 1, 2, \ldots, m \text{ (energy sectors)}$$
$$j = 1, 2, \ldots, n \text{ (consuming sectors)}$$

$$(\text{A.1})$$

This process allows us to project easily a change in energy flows in the economy prompted by a change in final demand by (1) obtaining the projected 'change-in-transactions' matrix (ΔX) discussed in chapter 2 and then (2) multiplying (element-by-element) the energy rows of the ΔX matrix (ΔE^e) by the table of inverse prices (**P**):

$$\Delta Z = (I - A)^{-1} \Delta Y \qquad\qquad (\text{A.2})$$
$$\Delta Z = A \Delta \hat{X}$$
$$\Delta Z^e = \text{the } m \text{ energy rows of } \Delta Z$$
$$\Delta E_{kj} = Z^e_{kj} \cdot P_{kj}$$

where $\Delta \mathbf{E}$ is the resulting change in energy flows

ΔE_{kj}^{e} is an element of $\Delta \mathbf{E}^{e}$

P_{kj} is an element of $\mathbf{P}$.

This notion of implied inverse energy prices makes it quite easy to develop a direct energy requirements matrix using the dollar-based technical coefficients matrix. Let us define $\mathbf{D}$ to be the direct energy requirements matrix. $\mathbf{D}$ is of dimension $m \times n$ and is found as follows:

$$D_{kj}(\text{Btu/\$}) = P_{kj}(\text{Btu/\$}) \cdot A_{kj}^{e}(\text{\$/\$}) \tag{A.3}$$

where P_{kj} is an element of $\mathbf{P}$, the implied inverse energy price matrix; and A_{kj}^{e} is an element of $\mathbf{A}^{e}$, the energy rows of the technical coefficients matrix.

Existence of $\mathbf{D}$ provides an alternative for calculating the energy flows $(\mathbf{X})$:

$$E_{kj} = D_{kj} \cdot Z_{j} \tag{A.4}$$

where Z_{j} is the total output (\$) of industry j.

Actually this is equivalent to equation (A.3) for calculating the energy flows since:

$$E_{kj} = P_{kj} \cdot X_{kj}$$

but $\quad X_{kj} [\mathbf{A}\hat{\mathbf{X}}^{-1}]_{kj} = A_{kj} \cdot X_{kj}$

and $\quad D_{kj} = P_{kj} \cdot A_{kj}$

so $\quad E_{kj} = D_{kj} \cdot Z_{j}$

The alternative is presented only because in some applications the available data favors one over the other.

We should note that P_{kj} is the amount of energy (in Btu's of the type supplied by energy sector k) associated with a dollar transaction from industry k to industry j. D_{kj} is the amount of energy (again in Btu's of the type supplied by energy sector k) but required to deliver a dollar's worth of industry j's output.

Finally, we wish to determine the total (indirect and direct) energy requirement needed to support some specified final demand. Let us first recall that:

$$AX + Y = X$$

or $\quad X = (I - A)^{-1}Y$

where $(I - A)^{-1}$ is the matrix of total dollar transactions required to support the final demand Y.

We can convert $(I - A)^{-1}$ to energy terms simply by multiplying by D (our direct transformation from dollars to Btu's). However, we may also wish to include energy sold directly to final demand. Let us define q_k to be the price (actually another implied inverse price) of energy type k sold directly to final demand, and S to be the $m \times n$ matrix made up of all q_k's. Note that there can only be m non-zero entries in S since we have assumed that only energy sectors can sell energy either to industry or final demand. Hence, S has elements:

$$S_{ki} = \begin{cases} q_k, & i = k \\ 0, & i \neq k \end{cases}$$

Then, accounting for energy sales to both industry and final demand, we can construct the total energy requirements by the following:

$$e = D(I - A)^{-1} + S$$

Hence, e_{kj}, an element of e, then describes the total amount of energy of type k required (directly and indirectly) to support a dollar's worth of industry i's output. Of course, it follows that:

$$F = eY$$

F is the vector of total energy outputs (in Btu's) required to support final demand Y (in $).

Employment and pollution intensities.
Several studies have developed schemes for relating levels of employment, pollution and other effects to the total requirements matrix with the result being a 'total employment requirements' matrix or 'total pollution generation requirements' much like the total energy requirements matrix just

described. In particular, vectors such as **w** (employment per dollar of total output – employment coefficients) or **v** (pollution of a particular type per dollar of total output – pollution coefficients) allows one to construct intensity matrices (see Folk and Hannon [130]):

$$\mathbf{v}(\mathbf{I} - \mathbf{A})^{-1} = \text{pollution intensity}$$
$$\mathbf{w}(\mathbf{I} - \mathbf{A})^{-1} = \text{employment intensity}$$

In this case, we can concatenate the energy coefficients, pollution coefficients, and employment coefficients to result in a total intensity matrix:

$$(\mathbf{I} - \mathbf{A})^{-1}\mathbf{Y} = \mathbf{X} \qquad \text{(The Leontief Inverse)}$$

$\mathbf{B}\,\mathbf{X} = \mathbf{X}^*$, $\mathbf{B} = $ the matrix of energy, environmental and employment coefficients (per unit of output)

$$\mathbf{B} = \begin{bmatrix} \mathbf{u} \\ \hline \mathbf{v} \\ \hline \mathbf{w} \end{bmatrix} \quad \begin{array}{l} \text{energy coefficients} \\ \text{environmental coefficients} \\ \text{employment coefficients} \end{array}$$

$$\mathbf{B}^* = \mathbf{B}\,(\mathbf{I} - \mathbf{A})^{-1}$$
$$\mathbf{X}^* = \mathbf{B}^*\,\mathbf{Y}$$

The total impact $\mathscr{X} = (\mathbf{X}, \mathbf{X}^*)$ is found by:

$$\mathscr{X} = \mathbf{T}^*\,\mathbf{Y} \qquad \mathbf{T}^* = \begin{bmatrix} (\mathbf{I} - \mathbf{A})^{-1} \\ \hline \mathbf{B}^* \end{bmatrix}, \text{ the total impact matrix}$$

Other methods of incorporating environmental effects, both pollution generation and abatement, into an input-output framework have been developed (Miernyk[153], Victor[169], and others). A pollution sector's row in an input-output table could reflect negative inputs (outputs) while the column would reflect negative outputs, e.g. abatement or recycling.

Appendix B

The RAS modification technique

The RAS method appears to have been formulated by Leontief himself and later developed by Stone and Brown. The most rigorous treatment of RAS, in a theoretical framework, appears to be that of Bacharach [21].

Modification procedure
To construct the new table (target table) using the RAS technique, the following information about the table to be derived must either be projected or observed:

1. Total gross outputs;
2. Total intermediate output – or total sales to final demand since gross outputs minus final demand equals intermediate output;
3. Total intermediate inputs – or the total value added since gross inputs (outputs) minus the value added equals intermediate input.

Let us consider the Direct Requirements matrix of the existing table to be $A(0)$ and that of the table to be derived as $A(1)$. Hence, it follows from above that all of the following are known:

$$X(1) - \text{Gross outputs};$$
$$Y(1) - \text{Sales to final demand};$$
$$VA(1) - \text{Value added};$$
$$U(1) - \text{Intermediate outputs};$$
$$V(1) - \text{Intermediate inputs};$$
$$A(0) - \text{The base technical coefficients matrix}$$

If the $A(1)$ matrix were known, the following relations would hold:

$$\left. \begin{array}{l} A(1)X(1) = U(1) \\ iA(1)\hat{X}(1) = V(1) \end{array} \right\} \qquad \text{where } \hat{X} = \begin{bmatrix} X_1 & \cdots & 0 \\ & X_2 & & \\ & & \ddots & \\ & & & \ddots \\ 0 & & \cdots & X_n \end{bmatrix}$$

and $\mathbf{i}$ is a $1 \times n$ identity vector.

However, since the $\mathbf{A}(1)$ is not known, we wish to find the proper a_{ij}'s – elements of $\mathbf{A}(1)$ to satisfy these equations.

As a first approximation, the technical coefficients of the existing I/O table can be used:

$$\mathbf{A}(0) \, \mathbf{X}(1) = \mathbf{U}^1 \tag{B.1}$$
$$\mathbf{i}\mathbf{A}(0) \, \hat{\mathbf{X}}(1) = \mathbf{V}^1 \tag{B.2}$$

where $\mathbf{U}^1$ and $\mathbf{V}^1$ are a first estimate of $\mathbf{U}(1)$ and $\mathbf{V}(1)$. In order to make the rows of the related transactions matrix, i.e. $\mathbf{A}(0) \, \hat{\mathbf{Z}}(1) = \mathbf{X}^1$, sum to the current outputs $\{\mathbf{U}(1)\}$ we can multiply each row of $\mathbf{A}(0)$ by:

$$r_i = \frac{u_i(1)}{u_i^1}$$

This operation forces $\mathbf{U}^1$ to be equal to $\mathbf{U}(1)$.

This new technical coefficients matrix, which we can call $\underline{\mathbf{A}}^1$, has elements:

$$d_{ij}' = a_{ij} \frac{u_i(1)}{u_i^1};$$
$$\mathbf{A}^1(1) \, \mathbf{X}(1) = \mathbf{U}(1)$$

The row adjustment is written as follows:

$$\mathbf{A}^1 = \hat{\mathbf{U}}(1)(\hat{\mathbf{U}}^1)^{-1}\mathbf{A}(0)$$

In order for this adjustment process to be useful in modifying the $\mathbf{A}$ matrix in a meaningful way, we must also adjust so that the columns of the new matrix sum to current inputs. It will generally be true that:

$$\mathbf{A}^1 \, \mathbf{X} \neq \mathbf{V}(1)$$

The corresponding operation for column-wise adjustment of the $\mathbf{A}$ matrix (we can operate on the previously row adjusted matrix $\mathbf{A}^1$) is:

$$\mathbf{A}^2 = \mathbf{A}^1 \, \hat{\mathbf{V}}(1)(\hat{\mathbf{V}}^1)^{-1}, \text{ where } \mathbf{A}^2 \text{ is a second estimate of } \mathbf{A}(1).$$

Note that for column-wise operations we must post-multiply the **A** matrix by the adjustment factor. The adjustment factor forces the columns of the resultant transactions matrix to sum to current inputs:

$$s_j = \frac{V_j(1)}{V_j^1}$$

$$\mathbf{i}\,\mathbf{A}^1\mathbf{s}\,\mathbf{X}(1) = \mathbf{V}(1)$$

$$\mathbf{i}\,\mathbf{A}^2\,\mathbf{X}(1) = \mathbf{V}(1)$$

The row-wise adjustment aberrates the column-wise sums from current inputs while the column-wise adjustment does the opposite, i.e. aberrates the row-sums from current outputs. Bacharach[121] proves that successive adjustment (iteration) converges. The convergence test is simply the successive estimates of current inputs and outputs (U^i and V^i) measured against the known values of U(1) and V(1).

We can note that by convention the adjustment factors have been termed r and s, hence, the name 'RAS' adjustment technique. The pre-multiplying row-wise adjustment factor is r and the post-multiplying column-wise adjustment factor is s. Successive estimates of the adjusted I/O table are found by:

$$\mathbf{A}^i = \mathbf{r}_i\,\mathbf{A}^{i-1}\,\mathbf{s}_i$$

Bibliography

PLANNING

1. Arrow, K. J., *Social Choice and Individual Values*, John Wiley and Sons, New York, 1963.
2. Baumol, William J., *Economic Theory and Operations Analysis*, 3rd ed., Prentice-Hall, Englewood Cliffs, New Jersey, 1973.
3. Chinitz, Benjamin, 'Appropriate Goals for Regional Economic Policy,' in D. L. McKeen, R. Dean and W. H. Leahy (eds.), *Regional Economics: Theory and Practice*, pp. 221–228, The Free Press, New York, 1970.
4. Davidoff, Paul and Thomas Reiner, 'A Choice Theory of Planning,' *Journal of the American Institute of Planners*, Vol. 28, No. 2, pp. 103–115, May 1962.
5. Denton, J. C., 'Technology Transform Analysis,' University of Pennsylvania, October 1973.
6. Denton, J. C., T. Saaty, P. Blair, F. Ma and P. Buneman, 'Planning for a Program Design for Energy Environmental Analysis,' The Energy Center, University of Pennsylvania, March 1976.
7. Dyckman, John W., 'Planning and Decision Theory,' *Journal of the American Institute of Planners*, Vol. 27, No. 4, pp. 335–345, November 1961.
8. Hill, M., *Planning for Multiple Objectives*, Regional Science Research Institute, Monograph Series, No. 5, Philadelphia, 1973.
9. Hirsch, Werner Z. (ed.), *Regional Accounts for Policy Decisions*, Johns Hopkins University Press, Baltimore, 1966.
10. Kahn, A. J., *Theory and Practice of Social Planning*, Chapter IV, Russel Sage Foundation, New York, 1969.
11. Land, Kenneth, 'Theories, Models and Indicators of Social Change,' *Journal of International Social Science*, Vol. 27, No. 1, 1975.
12. Lewin, A. Y. and M. F. Shakun, *Policy Sciences: Methodologies and Cases*, Pergamon Press, New York, 1975.
13. Ozbekhan, Hasan, 'Planning and Human Action,' University of Pennsylvania, 1973.
14. Ozbekhan, Hasan, 'Thoughts on the Emerging Methodology of Planning,' Wharton School, University of Pennsylvania, 1975.
15. Reiner, Thomas, 'A Multiple Goals Framework for Regional Planning,' *Regional Science Association Papers*, Vol. 26, pp. 207–239, 1971.
16. Rondinelli, D. A., 'Planning and Development Policy,' *Urban Affairs Quarterly*, pp. 30–39, September 1971.
17. Salih, Kamal, 'Goal Conflicts in Pluralistic Multi-Level Planning for Development,' *International Regional Science Review*, Vol. 1, No. 1, Spring 1975.
18. Vickers, Geoffry, *Value Systems and Social Process*, Basic Books, New York, 1968.

19. Vickers, Geoffry, 'Values, Norms and Policies,' *Policy Sciences*, Vol. 4, No. 1, pp. 103–111, March 1973.
20. Vickers, Geoffry, *The Art of Judgment*, Basic Books, New York, 1965.

SYSTEMS THEORY

21. Ackoff, R. L. and F. E. Emery, *On Purposeful Systems*, Aldine-Atherton, Chicago, 1972.
22. The ADAR Corporation, 'The Sudan Transport Study,' Vol. 1–5, Philadelphia, October 1974.
23. Ashby, W. R., *An Introduction to Cybernetics*, John Wiley and Sons, New York, 1958.
24. Von Bertolanffy, Ludwig, *General Systems Theory*, George Braziller, New York, 1968.
25. Von Bertolanffy, Ludwig, *Perspectives on General Systems Theory*, George Braziller, New York, 1975.
26. Boulding, Kenneth, 'Economics and General Systems,' *International Journal of General Systems*, Vol. 1, No. 1, pp. 67–73, 1974.
27. Boulding, Kenneth, 'General Systems Theory, the Skeleton of Science', *Management Science*, Vol. 2, No. 3, pp. 197–208, April 1956.
28. Brown, Seyom, 'Scenarios in Systems Analysis,' *Systems Analysis and Policy Planning: Applications in Defense*, E. S. Quade and W. I. Boucher (eds.) American Elsevier Publishing Company, New York, 1969.
29. Churchmann, C. W., *Design of Inquiring Systems: Basic Concepts in Systems Analysis*, Basic Books, New York, 1972.
30. Churchman, C. W., *The Systems Approach*, Dell Publishing Company, New York, 1968.
31. *The Economist*, p. 103, May 10, 1975.
32. Emery, F. E., *Systems Thinking*, Penguin Books, New York, 1970.
33. Gantmacher, F. R., *The Theory of Matrices*, Vol. 2, pp. 53 and 60, Chelsea, New York, 1960.
34. Householder, A. S., *The Theory of Matrices in Numerical Analysis*, Dover Publications, New York, 1964.
35. Klir, George, *Trends in General Systems Theory*, Wiley-Interscience, New York, 1972.
36. Kraemer, Kenneth, *A Systems Approach to Decision Making*, International City Management Association, Washington, D.C., 1973.
37. Laslo, Ervin, *Introduction to Systems Philosophy*, Gordon and Breach, New York, 1972.
38. Laslo, Ervin, *The Relevance of General Systems Theory*, George Braziller, New York, 1972.
39. Laslo, Ervin, 'A System Philosophy of Human Values,' *Behavioral Science*, Vol. 18, No. 4, pp. 250–259, July 1973.
40. Laslo, Ervin, *The Systems View of the World*, George Braziller, New York, 1972.
41. Lewin, A. Y. and M. F. Shakun, 'Situational Normativism: A Descriptive Approach to Decision Making and Policy Sciences,' *Policy Sciences*, March 1976.
42. Mariano, R. S., 'Allocation Models for Energy Planning,' (unpublished Ph.D. dissertation), University of Pennsylvania, 1975.
43. Mesarovic, M. D. and D. Macko, 'Scientific Theory of Hierarchical Systems,' in *Hierarchical Structures*, L. L. Whyte, A. G. Wilson, D. Wilson (eds.), American Elsevier Publishing Company, New York, 1969.
44. Mesarovic, M. D., D. Macko and Y. Takahara, *Theory of Hierarchical Multilevel Systems*, Academic Press, New York, 1970.
45. Mihram, G. A., *Simulation: Statistical Foundations and Methodology*, Academic Press, New York, 1972.
46. Miller, G. A., 'The Magical Number Seven Plus or Minus Two: Some Limits on Our Capacity for Processing Information,' *The Psychological Review*, Vol. 63, pp. 81–97, March 1956.

47. Pattee, H. H. (ed.), *Hierarchy Theory*, George Braziller, New York, 1973.
48. Saaty, T. L., *Mathematical Models of Arms Control and Disarmament – Application of Mathematical Structures in Politics*, John Wiley, New York.
49. Saaty, T. L., 'Scaling Method for Priorities in Hierarchical Structures,' *Journal of Mathematical Psychology*, Vol. 15, No. 3, pp. 234–281, June 1977.
50. Saaty, T. L., 'Structure, Flow and Purpose in Systems,' (unpublished), University of Pennsylvania, 1976.
51. Saaty, T. L., *Topics in Behavioral Mathematics*, Mathematical Association of America, 1973.
52. Saaty, T. L. and P. C. Rogers, 'The Future of Higher Education in the U.S. (1985–2000),' *Socio-Economic Planning Sciences*, Vol. 10, No. 6, pp. 251–264, December 1976.
53. Saaty, T. L. and M. Khouja, 'A Measure of World Influence,' *Journal of Peace Science*, No. 1, 1976.
54. Shakun, M. F., 'Policy Making Under Discontinuous Change: The Situational Normativism Approach,' *Management Science*, Vol. 22, No. 2, October 1975.
55. Sutherland, John, *A General Systems Philosophy for the Social and Behavioral Sciences*, George Braziller, New York, 1973.
56. Weinberg, G., *An Introduction to General System Theory*, Wiley-Interscience, New York, 1975.
57. Weiss, P. A., *Hierarchically Organized Systems*, Hafner, 1971.
58. Whyte, L. L., A. G. Wilson and D. Wilson, *Hierarchical Structures*, American Elsevier Publishing Company, New York, 1969.
59. Wielandt, H., 'Unzerlegbare, nicht negative Matrizen,' *Mathematische Zeitschrift*, Vol. 52, pp. 642–648, 1950.
60. Wilkinson, J. J., *The Algebraic Eigenvalue Problem*, Oxford University Press, London, 1965, Chapter 2.

ELECTRIC POWER AND ENERGY PARKS

61. Blair, P., F. Rahbar, D. Hu, R. Ziokowski and L. Smith, 'Development of the Energy Park Concept in Pennsylvania – the Montour Site,' (unpublished), The Energy Center, University of Pennsylvania, May 7, 1975.
62. Brookhaven National Laboratory, 'Preliminary Assessment of a Hypothetical Nuclear Energy Center in New Jersey,' Energy Policy Analysis Group, Upton, New York, November 1975.
63. Cooke, M. L. (ed.), *Giant Power: Large Scale Electrical Development as a Social Factor*, Annals of the American Academy of Political and Social Science, Philadelphia, 1925.
64. DuBoff, R. B., 'Electric Power in American Manufacturing 1889–1958,' (unpublished Ph.D. dissertation), University of Pennsylvania, 1964.
65. 'Electric Power and Government Policy,' Twentieth Century Fund, Power committee, New York, 1948.
66. Federal Power Commission, 'The 1970 National Power Survey,' U.S. Government Printing Office, Washington, D.C., 1972.
67. Ferrar, Terry A., 'Energy Parks – Report to Pennsylvania Governor's Energy Council,' Philadelphia, Pennsylvania, October 1975.
68. General Electric Company, 'Assessment of Energy Parks vs. Dispersed Electric Power Generating Facilities,' Center for Engery Systems, (interim report to NSF), CFES-PR-74-5, October 1974.
69. Gilbert Associates, Inc., 'Energy Park Development Group Site Selection Study,' GAI-1853, Vols. I and II, Reading, Pennsylvania, January 1975.

70. Hunt, E. E., 'The Power Industry and the Public Interest,' (unpublished Ph.D. dissertation), University of Pennsylvania, February 1944.
71. Isard, W., T. Reiner, R. Van Zele and J. Strathman, 'Regional Economic Impacts of Nuclear Power Plants,' University of Pennsylvania, Department of Regional Science, 1976.
72. Lovins, A. D. and J. H. Price, *Non-Nuclear Futures*, Ballinger, Cambridge, Mass., 1975.
73. MacLaren, Malcolm, *The Rise of the Electrical Industry During the 19th Century*, Princeton University Press, Princeton, 1943.
74. Meier, R. L., 'The Social Impact of a Nuplex,' *Bulletin of Atomic Scientists*, Vol. 25, pp. 16–21, March 1969.
75. Meier, Peter M. and Philip F. Palmedo, 'Planning Nuclear Energy Centers Under Technological and Demand Uncertainty,' Brookhaven National Laboratory, presented at ORSA/TIMS 1976 Joint National Meeting, Philadelphia, April 2, 1976.
76. Mulford, R., Philadelphia Electric Company, (personal communication), January 1976.
77. Murray, W. S., *Superpower, its Genesis and Future*, McGraw-Hill, New York, 1925.
78. National Electric Reliability Council, 'Nuclear Energy Centers – An Assessment of Impact on Reliability of Electric Power Supply,' April 1975.
79. Novick, S., 'The Electric Power Industry,' *Environment*, Vol. 17, No. 8, November 1975.
80. 'Nuclear Standards and Standard Nuclear Plants: More than Money at Stake,' *Power Engineering*, March 1975.
81. Oak Ridge National Laboratory, 'Analysis of the Gulf States Utilities River Bend Site for a Nuclear Energy Center,' ORNL CF-74-1-32, Oak Ridge, Tennessee, January 1974.
82. Oak Ridge National Laboratory, 'General Report on the Middle East Study Project,' ORNL-4481, Oak Ridge, Tennessee, September, 1970.
83. Oak Ridge National Laboratory, 'Nuclear Energy Centers – Industrial and Agro-Industrial Complexes,' ORNL-4290, ORNL-4291, Oak Ridge, Tennessee, November 1968.
84. Peele, Elizabeth, 'Social Effects of Nuclear Power Plants' in C. P. Wolf (ed.) *Social Impact Assessment*, Proceedings of Environmental Design and Research Association Conference, Milwaukee, Wisconsin, 1974.
85. Pennsylvania Energy Park Development Group, 'Energy Parks in Pennsylvania,' January 1975.
86. Pennsylvania Energy Park Development Group, 'Transmission Task Force Report,' January 1975.
87. Pennsylvania State University, 'Energy Parks and the Commonwealth of Pennsylvania – Issues and Recommendations,' Vol. I and II, Center for the Study of Environmental Policy, July 1975.
88. Rasmussen, N. C., 'An Assessment of Accident Risks in U.S. Commercial Nuclear Power Plants,' WASH-1400 (Draft), U.S. Atomic Energy Commission, August 1974
89. Reiner, T., J. Strathman and R. Van Zele, 'Regional Economic Impacts of Nuclear Energy Centers,' University of Pennsylvania, Department of Regional Science, 1975.
90. Riorden, Courtney, 'A Study of the Probable Economic and Social Effects of Constructing and Operating the Bell Nuclear Power Station in Tompkins County,' Cornell University, Office of Regional Resources, Ithaca, New York, 1970.
91. Rombaugh, C. and B. V. Koen, 'Total Energy Investment in Nuclear Power Plants,' *Nuclear Technology*, Vol. 26, pp. 5–11, May 1975.
92. Saeger, John H., 'Pennsylvania Energy Park Study,' Pennsylvania Energy Park Development Group, presented to the Pennsylvania Electric Association System Planning Committee, Wilkes Barre, Pennsylvania, October 21, 1975.
93. Schuller, C. R., 'Citizen's Views About the Proposed Hartsville Nuclear Power Plant: a Preliminary Report of Potential Social Impacts,' Oak Ridge National Laboratory, Oak Ridge, Tennessee, 1975.

94. Sharko, J. R. and D. R. Limaye, 'Quantitative Models in the Energy Sector: a Review of the State of the Art,' Decision Sciences Corporation Research Paper Series, No. 114, Jenkintown, Pennsylvania, 1973.

95. Shoemaker, Mary F., 'The Impact of Salem Nuclear Generating on Lower Alloways Creek Township,' unpublished M.A. thesis, Glassboro State College, 1975.

96. Smiley, S. H., M. L. Ernst, G. Sege and R. T. Jaske, 'Feeding the Glutton,' Spectrum, Vol. 13, No. 7, pp. 74ff, July 1976.

97. Soldat, J. K., 'Radiological Impact of a Nuclear Energy Center on the Environment,' Battelle Northwest Laboratory, BNWL-B 333, Richland, Washington, January 1974.

98. Sporn, Philip, Vistas in Electric Power, Pergamon, New York, 1968.

99. United Engineers and Constructors, Inc., 'Construction City Study for a Nuclear Power Center,' Report to the U.S. Atomic Energy Commission (draft), January 1974.

100. University of Pennsylvania, 'Regional Economic Impacts of Nuclear Energy Centers,' Regional Science Department, Philadelphia, Pennsylvania, 1975.

101. U.S. Atomic Energy Commission, '1000MW Control Station Power Plants Investment Cost Study,' WASH-1230, Vol. 1, 1971.

102. U.S. Atomic Energy Commission, 'Evaluation of Nuclear Energy Centers,' Office of Reactor Research, WASH-1288, January 1974.

103. U.S. Atomic Energy Commission, 'Land Use and Nuclear Power Plants,' WASH-1319, 1974.

104. U.S. Atomic Energy Commission, 'Power Parks, A Preliminary Analysis,' Energy Policy Branch, Office of Planning and Analysis, September 1973.

105. U.S. Atomic Energy Commission, 'Power Plant Capital Costs: Current Trends and Sensitivity to Economic Parameters,' WASH-1345, 1974.

106. U.S. Atomic Energy Commission, 'Projections of Labor Requirements for Electrical Power Plants 1974 – 2000,' WASH-1334, 1974.

107. U.S. Atomic Energy Commission, 'Puerto Rico Energy Center Study,' TID-25602, July 1970.

108. U.S. Code, 42 SCU 5847, Section 207, 'Nuclear Energy Center Site Survey.'

109. U.S. Executive Office of the President, 'Considerations Affecting Steam Power Plant Site Selection,' Energy Policy Office, Office of Science and Technology, Washington, D.C., December 1968.

110. U.S. Nuclear Regulatory Commission, 'Nuclear Energy Center Site Survey – 1975,' NUREG-0001, Vols. I–V, Washington, D.C., 1976.

111. U.S. Nuclear Regulatory Commission, 'Nuclear Energy Center Site Survey,' Office of Special Studies, NUREG-75/018M, March 1975.

112. U.S. Nuclear Regulatory Commission, 'Conceptual Description of Nuclear Energy Centers,' Nuclear Energy Center Site Survey, May 1975.

113. U.S. Nuclear Regulatory Commission, 'Guidelines for Workshop Discussions of Practical Issues of Implementation of Nuclear Energy Centers,' Office of Special Studies, May 1975.

114. U.S. Nuclear Regulatory Commission, 'Nuclear Energy Center Site Survey – Practicality Workshops – EAST,' MITRE Corporation, MTR-6967, July 1975.

115. U.S. Nuclear Regulatory Commission, 'Nuclear Energy Centers Survey – Practicality Workshops – WEST,' Rand Corporation, May 1975.

116. Weinberg, A. M., 'Demographic Policy and Power Plant Siting,' (submitted to the U.S. Senate Committee on Internal and Insular Affairs), October 1971.

117. Weinberg, A. M., 'Energy Needs, Nuclear Power and the Environment,' Fifth International Congress of Radiation Research, Seattle, Washington, October 1973.

118. Weinberg, A. M., 'On Siting of Nuclear Reactors,' U.S. Atomic Energy Commission, February 1972.

119. Weinberg, A. M., 'The Moral Imperatives of Nuclear Energy,' Nuclear News, pp. 33–37, December 1971.

120. Weinberg, A. M. and C. C. Burwell, 'A Strategy for Research and Development, A Position Paper,' Oak Ridge National Laboratory, October 1973.

INPUT-OUTPUT AND IMPACT ANALYSIS

121. Bacharach, M., *Biproportional Matrices and Input-Output Change*, Cambridge University Press, Cambridge, 1970.
122. Baughman, Martin L., 'Energy System Modeling Regulation and New Technology' in Michael S. Macrakis (ed.) *Energy Demand, Conservation and Institutional Problems*, MIT Press, Massachusetts, 1974.
123. Baumol, William J., *Economic Theory and Operations Analysis*, 3rd ed., Prentice-Hall, Englewood Cliffs, 1972.
124. Bezdek, R. and B. Hannon, 'Derivation of the 1963 and 1967 Total Employment Vector for 362 I/O Sectors,' CAC Document No. 63, Center for Advanced Computation, University of Illinois at Urbana-Champaign, April 1973.
125. Blair, P., 'Nonsurvey Estimates of Input-Output Tables,' (unpublished), University of Pennsylvania, February 1975.
126. Dantzig, G., *Linear Programming and Extensions*, Princeton University Press, Princeton, N.J., 1963.
127. Dorfman, R., P. Samuelson and R. Solow, *Linear Programming and Economic Analysis*, McGraw-Hill, New York, 1958.
128. Dupree, W. and O. R. West, *United States Energy Through 2000*, U.S. Department of the Interior, Washington, D.C., December 1972.
129. Federal Energy Administration, *Project Independence Report*, U.S. Government Printing Office, Washington, D.C., November 1974.
130. Folk, H. and B. Hannon, 'An Energy, Pollution and Employment Policy Model' in M. S. Macrakis (ed.), *Energy Demand, Conservation and Institutional Problems* MIT Press, Massachusetts, 1974.
131. Ford, Andrew, 'Environmental Policies for Electricity Generation: a Study of the Long-Term Dynamics of the SO Problem,' *Energy Systems and Policy*, Vol. 1, No. 3, March 1976.
132. Ford Energy Policy Project, 'A Time to Choose: America's Energy Future,' Ballinger, Cambridge, Mass., 1974.
133. Gamble, H. B. and D. L. Raphael, 'Microregional Analysis of Clinton County,' Pennsylvania State University, 1966.
134. Herendeen, Robert A., 'An Energy Input Output Matrix for the United States, 1963: User's Guide,' Document No. 69. Center for Advanced Computation, University of Illinois at Urbana-Champaign, March 1973.
135. Herendeen, Robert A., 'The Energy Cost of Goods and Services,' Environmental Report, ORNL-NSF Environmental Program, Oak Ridge National Laboratory, Oak Ridge, Tennessee, 1974.
136. Herendeen, Robert A. and C. W. Bullard III, 'Energy Cost of Goods and Services, 1963 and 1967,' CAC Document No. 140, Center for Advanced Computation, University of Illinois at Urbana-Champaign, November 1974.
137. Hewings, G. J., 'Regional Input-Output Models in the United Kingdom: Some Problems and Prospects for the Use of Nonsurvey Techniques,' *Regional Studies*, Vol. 5, 1971.
138. Hoffman, K. C., 'A Unified Framework for Energy System Planning,' *Energy Modeling*, Resources for the Future, Inc., Washington, D.C., 1973.
139. Hudson, E. and D. W. Jorgenson, 'The U.S. Energy Policy and Economic Growth 1975–2000,' *The Bell Journal of Economics and Management Science*, Vol. 5, No. 2, 1974.

140. 'Input-Output Structure of the U.S. Economy: 1967,' *Survey of Current Business*, February 1974.
141. Intrilligator, M., *Mathematical Optimization and Economic Theory*, Prentice-Hall, Englewood Cliffs, N.J., 1971.
142. Isard, W., *Methods of Regional Analysis, An Introduction to Regional Science*, MIT Press, New York, 1960.
143. Isard, W. and T. Langford, *Philadelphia Regional Input-Output Study*, MIT Press, Massachusetts, 1971.
144. Istvan, R., 'Interindustry Impacts of Alternative Utility Investment Strategies' in M. Macrakis (ed.), *Energy Demand, Conservation and Institutional Problems*, MIT Press, Massachusetts, 1974.
145. Jorgenson, Dale, 'An Integrated Reference Energy System and Inter-Industry Model for the United States Economy,' in J. A. Tomlin (ed.), *Notes on a Workshop on Energy Systems Modeling*, Systems Optimization Laboratory, Stanford University, April 1975.
146. Just, J., 'Changes in Energy Consumption 1963-1980,' (unpublished Master's thesis), MIT, Massachusetts, 1973.
147. Langford, T. and W. Isard, 'Philadelphia Region Input-Output Study-Working Papers,' University of Pennsylvania, Department of Regional Science, 1967.
148. Leontief, W., 'Environmental Repercussions and the Economic Structure: An Input-Output Approach,' *Review of Economic Statistics*, Vol. 52, pp. 262–271, August 1970.
149. Leontief, W., *The Structure of the American Economy 1919–1939*, Oxford University Press, New York, 1951.
150. Malizia and D. L. Bond, 'Empirical tests of the RAS Method of Coefficient Adjustment,' *Journal of Regional Science*, Vol. 14, No. 3, December 1974.
151. McMenamin, D. and J. Haring, 'An Appraisal of Nonsurvey Techniques for Estimating Regional Input-Output Models,' *Journal of Regional Science*, Vol. 14, August 1974.
152. Mesarovic, M. and E. Pistel, *Mankind at the Turning Point, Second Report to the Club of Rome*, E. P. Dutton, New York, 1974.
153. Miernyk, W. H., *Elements of Input-Output Economics*, Random House, New York, 1965.
154. Miller, R. E., 'Foundations of Input-Output Analysis,' (unpublished) University of Pennsylvania, Department of Regional Science, 1974.
155. Morrison, W. I. and P. Smith, 'Nonsurvey Input-Output Techniques at the Small Area Level: An Evaluation,' *Journal of Regional Science*, Vol. 14, April 1974.
156. National Petroleum Council, 'U.S. Energy Outlook,' Washington, D.C., December 1972.
157. Palmedo, P. F. (ed), 'Brookhaven Regional Energy Studies Program Annual Report, 1975,' Brookhaven National Laboratory, Department of Applied Science, November 1975.
158. Pennsylvania Department of Commerce, 'County Industry Report,' Bureau of Economic Analysis, 1974.
159. Reardon, W., 'Input-Output Analysis of U.S. Energy Consumption,' *Energy Modeling*, Resources for the Future, Inc., Washington, D.C., March 1973.
160. Saaty, T. L., F. Ma and P. Blair, 'A Regional Energy Environment Game with Actor Participation,' Paper 549, TIMS/ORSA National Meeting, Philadelphia, Pennsylvania, March 30 – April 2, 1976.
161. Schaffer, W. A., 'Estimating Regional Input-Output Coefficients,' *Review of Regional Studies*, Vol. 2, 1972.
162. Stanford Research Institute, 'SRI-Gulf Energy Model: An Overview of Methodology,' SRI International Research and Development Report, Decision Analysis Group, Stanford, California, January 1975.

163. Strout, A. M., 'Technological Change and United States Energy Consumption, 1939–1954,' (unpublished Master's thesis), University of Chicago, 1967.
164. Toscas, J., 'U.S. Air Pollution Generation in 1967,' CAC Technical Memorandum 15, Center for Advanced Computation, University of Illinois at Urbana-Champaign, January 8, 1974.
165. U.S. Bureau of Labor Statistics, 'Consumption Patterns in the Northeast,' Washington, D.C., 1974.
166. U.S. Bureau of Mines, 'Supply and Demand for Energy in the United States and Region, 1960 through 1965,' Bureau of Mines Information Circular 8434, U.S. Department of the Interior, 1970.
167. U.S. Department of Commerce, 'Fuels and Energy Consumed,' U.S. Census of Manufacturers, MC72(SR)-6, 1973.
168. U.S. Energy Research and Development Administration, 'New Energy Technology Coefficients and Dynamic Energy Models,' ERDA-3, Vols. 1 and 2, 1975.
169. Victor, P. A., *Pollution: Economy and Environment*, University of Toronto Press, Toronto, 1972.
170. White, D., 'Goods and Services: An Input-Output Analysis,' *Energy Policy*, Vol. 2, No. 4, December 1974.
171. Yan, C., *Introduction to Input-Output Economics*, Holt, Rinehart and Winston, New York, 1969.

MULTIOBJECTIVE DECISION MAKING (1950–1976)

172. Athans, M. and H. P. Geering, 'Necessary and Sufficient Conditions for Differentiable Nonscalar-Valued Functions to Attain Extrema,' *IEEE Transactions*, Vol. AC-18, No. 2, 1973.
173. Awerback, S., J. Ecker and W. Wallace, 'A Note: Hidden Nonlinearities in the Application of Goal Programming,' *Management Science*, Vol. 23, No. 1, pp. 918–920, January 1976.
174. Benayoun, R., J. deMontgolfier, J. Tergny and O. Laritchen, 'Linear Programming with Multiple Objective Functions: Step Method (STEM),' *Mathematical Programming*, Vol. 1, pp. 366–375, 1971.
175. Charnes, A. and W. W. Cooper, *Management Models and Industrial Applications of Linear Programming*, Vols. 1 and 2, John Wiley and Sons, New York, 1961.
176. Charnes, A., W. W. Copper and R. O. Ferguson, 'Optimal Estimation of Executive Compensation by Linear Programming,' *Management Science*, Vol. 1, No. 2, January 1955.
177. Charnes, A., W. W. Cooper, D. Klingman and R. J. Nilhaus, 'Explicit Solutions in Convex Goal Programming,' *Management Science*, Vol. 22, No. 4, pp. 4v8–448, December 1975.
178. Charnes, A., W. W. Cooper and R. J. Nilhaus, 'Goal Programming Model for Manpower Planning,' Management Science Research Report No. 115, Carnegie Mellon University, August 1968.
179. Charnes, A., W. W. Cooper, R. J. Nilhaus and A. Stedry,' Static and Dynamic Assignment Models with Multiple Objectives and some Remarks on Organization Design,' *Management Science*, Vol. 4, No. 3, April 1969.
180. Charnes, A., W. W. Cooper, R. J. Nilhaus and A. Stedry, 'A Goal Programming Model for Media Planning,' *Management Science*, Vol. 14, No. 8, pp. 423–430, April 1968.
181. Cochrane, James L. and Milan Zeleny (eds.), *Multiple Criteria Decision Making*, University of South Carolina Press, South Carolina, 1973.
182. Cohon, J. L. and D. H. Marks, 'Multiobjective Screening Models and Water Resource Investment,' *Water Resources Research*, Vol. 9, No. 4, 1973.

183. Cohon, J. L. and D. H. Marks, 'A Review and Evaluation of Multiobjective Programming Techniques,' *Water Resources Research*, Vol. 11, No. 2, April 1975.
184. Contini, B., 'A Stochastic Approach to Goal Programming,' *Operations Research*, Vol. 16, No. 3, pp. 576–596, 1968.
185. Dantzig, G., *Linear Programming and Extensions*, Princeton University Press, Princeton, N.J., 1963.
186. Dorfman, R., P. Samuelson and R. Solow, *Linear Programming and Economic Analysis*, McGraw-Hill, New York, 1958.
187. Dyer, J. S., 'An Empirical Investigation of a Man-Machine Interactive Approach to the Solution of the Multiple Criteria Problem,' discussion paper, Western Management Institute, University of California, Los Angeles, June 1971.
188. Dyer, J. S., 'Interactive Goal Programming,' *Management Science*, Vol. 19, No. 1, pp. 62–70, September 1972.
189. Evans, J. P. and R. E. Steuer, 'A Revised Simplex Method for Linear Multiple Objective Programs,' *Mathematical Programming*, Vol. 5, pp. 54–72, 1973.
190. Fishburn, P., 'Bernoullian Utilities for Multiple-Factor Situations,' in Cochrane J. and M. Zeleny (eds.), *Multiple Criteria Decision Making*, University of South Carolina Press, South Carolina, 1973.
191. Gass, Saul, *Linear Programming*, McGraw-Hill, New York, 1958.
192. Gass, Saul and T. Saaty, 'Parametric Objective Function – Part II, Generalization,' *Journal of the Operations Research Society of America*, Vol. 3, 1955.
193. Gass, Saul and T. Saaty, 'The Computational Algorithm for the Parameter Objective Function,' *Naval Research Logistics Quarterly*, Vol. 2, pp. 39–45, 1955.
194. Geoffrion, A. M., 'Strictly Concave Parametric Programming,' *Management Science*, Vol. 15, No. 13, 1966.
195. Geoffrion, A. M., 'Proper Efficiency and the Theory of Vector Maximization,' *Journal of Mathematical Analysis*, Vol. 22, pp. 618–630, 1968.
196. Goldberg, I., *Methods of Real Analysis*, Xerox College Publishing Company, Massachusetts, 1964.
197. Gros, J. G., R. Avenhaus, J. Linnerooth, P. D. Pahmer and H. J. Otway, 'A Systems Analysis Approach to Nuclear Facility Siting,' *Behavioral Science*, Vol. 21, pp. 116–127, 1976.
198. Haimes, Y., W. A. Hall and H. T. Freedman, *Multiobjective Optimization in Water Resource Systems*, Elsevier Scientific Publishing Company, New York, 1975.
199. Hamlin, S. and W. Hamlin, 'The Use of Goal Programming in Evaluating Air Pollution Control Strategies,' University of Buffalo, School of Management, March 1976.
200. Himmelblau, D. M., *Applied Nonlinear Programming*, McGraw-Hill, New York, 1972.
201. Ijiri, V., *Management Goals for Accounting and Control*, Rand McNally, Chicago, 1965.
202. Intrilligator, M., *Mathematical Optimization and Economic Theory*, Prentice-Hall, Englewood Cliffs, N.J., 1971.
203. Jaaskelainen, V., 'A Goal Programming Model of Aggregate Production Planning,' *Swedish Journal of Economics*, No. 2, pp. 14–19, 1969.
204. Johnsen, Eric, *Studies in Multiobjective Decision Models*, Student Literature, Lund, 1968.
205. Kapur, K. C., 'Mathematical Methods of Optimization for Multiobjective Transportation Systems,' *Socio-Economic Planning Science*, Vol. 4, pp. 450–467, 1970.
206. Koopmans, T. C. (ed.), *Activity Analysis of Production and Allocation*, John Wiley and Sons, New York, 1951.
207. Koopmans, T. C. (ed.), 'Analysis of Production as an Efficient Combination of Activities,' *Activity Analysis of Production and Allocation*, T. C. Koopmans (ed.), John Wiley and Sons, New York, 1951.
208. Kornbluth, J., 'A Survey of Goal Programming,' *Omega*, Vol. 1, No. 2, pp. 193–205, 1973.
209. Kuhn, H. W. and A. W. Tucker, 'Nonlinear Programming,' *Proceedings – Second*

Berkeley Symposium on Mathematical Statistics and Probability, University of California Press, California, 1950.

210. Lane, M., 'Goal Programming and Satisficing Models in Economic Analysis,' (unpublished Ph.D. dissertation), University of Texas, Austin, 1970.

211. Lee, Sang, *Goal Programming for Decision Analysis*, Auerbach, Philadelphia, 1972.

212. Lee, Sang, 'Goal Programming for Decision Sciences of Multiple Objectives,' *Sloan Management Review*, Vol. 14, No. 2, pp. 11–24, 1973.

213. Lee, Sang, 'Decision Analysis Through Goal Programming,' *Decision Sciences*, Vol. 2, No. 2, pp. 172–180, 1971.

214. Lee, Sang and E. Clayton, 'A Goal Programming Model for Academic Resource Planning,' *Management Science*, Vol. 18, No. 8, pp. 395–408, April 1972.

215. Litchfield, J. W., J. V. Hansen and L. E. Beck, 'A Research and Development Decision Model Incorporating Utility Theory and Management of Social Values,' *IEEE Transactions*, Vol. SMC-6, No. 6, pp. 400ff, June 1976.

216. Loucks, D. P., 'Conflict and Choice: Planning for Multiple Objectives' in *Economy-Wide Models and Development Planning*, C. Blitzer, P. Clark and L. Taylor (eds.), Oxford University Press, New York, 1975.

217. Panagiotakopoulos, Demetrios, 'A Multi-Objective Framework for Environmental Management Using Goal Programming,' *Journal of Environmental Systems*, Vol. 5, No. 2, 1975.

218. Philip, Johan, 'Algorithms for the Vector Maximization Problem,' *Mathematical Programming*, Vol. 2, No. 2, pp. 207–229, April 1972.

219. Raiffa, H., *Decision Analysis*, Addison-Wesley, Massachusetts, 1968.

220. Reid, R. N. and S. J. Citron, 'On Non-Inferior Performance Index Vectors,' *Journal of Optimization Theory and Applications*, Vol. 7, No. 1, 1971.

221. Roy, Boothoyd, 'Problems and Methods with Multiple Objective Functions,' *Mathematical Programming*, Vol. 1, No. 2, pp. 239–266, November 1971.

222. Saaty, T. L. and S. Gass, 'The Parametric Objective Function – Part I,' *Journal of the Operations Research Society of America*, Vol. 2, 1954.

223. Sakawa, M. and Y. Sawaragi, 'Multiple-Criteria Optimization of Pollution Control Models,' *International Journal of Systems Science*, Vol. 6, No. 8, pp. 741–748, 1975.

224. Sakawa, M. and Y. Sawaragi, 'Multiple-Objective Optimization for Environmental Development Systems,' *International Journal of Systems Science*, Vol. 6, pp. 157ff, 1974.

225. Salukvadze, M. E., 'On the Existence of Solutions in Problems of Optimization Under Vector Valued Criteria,' *Journal of Optimization Theory and Applications*, Vol. 13, No. 2, 1974.

226. Savir, D., 'Multiobjective Linear Programming,' Report ORC 66 21, Operations Research Center, University of California, Berkeley, California, November 1966.

227. Spivey, W. A. and H. Tamura, 'Goal Programming in Econometrics,' *Naval Research Logistics Quarterly*, Vol. 17, No. 2, pp. 183–192, 1970.

228. Terry, H., 'Comparative Evaluation of Performance Using Multiple Criteria,' *Management Science*, Vol. 9, No. 3, pp. 431–442, 1963.

229. Wagner, H. M., *Principles of Operations Research*, Prentice-Hall, Englewood Cliffs, New Jersey, 1969.

230. Yu, P. L. and G. Leitmann, 'Compromise Solutions, Domination Structures and Salukvadze's Solution,' *Journal of Optimization Theory and Applications*, Vol. 13, No. 3, 1974.

231. Zadeh, A., 'Optimality and Non-Scalar Valued Performance Criteria,' *IEEE Transactions*, Vol. AC-8, No. 1, 1963.

232. Zoints, S. and J. Wallenius, 'An Interactive Programming Method for Solving the Multiple Criteria Problem,' *Management Science*, Vol. 22, No. 6, pp. 652–663, February 1976.

233. Zukhovitsky, S. I. and L. I. Avdyeva, *Linear and Convex Programming*, W. B. Saunders, Philadelphia, 1966.

MULTIOBJECTIVE DECISION MAKING: RECENT WORKS (1976–1977)

234. Alexander, J. and T. L. Saaty, 'The Forward and Backward Processes of Conflict Analysis,' *Behavioral Science*, March 1977.
235. Donckels, R., 'Regional Multiobjective Planning Under Uncertainty: A Stochastic Goal Programming Approach,' *Journal of Regional Science*, Vol. 17, No. 2, pp. 207–216, 1977.
236. Nijkamp, P., *Theory and Application of Environmental Economics*, North Holland, Amsterdam, 1977.
237. Nijkamp, P. and A. van Delft, *Multicriteria Analysis and Regional Decision-Making*, Martinus Nijhoff, Leiden, 1977.
238. Nijkamp, P. and A. van Delft, 'The Use of Hierarchical Optimization Criteria in Regional Planning,' *Journal of Regional Science*, Vol. 17, No. 2, pp. 195–205, 1977.
239. Nijkamp, P. and P. Rietveld, 'Multiobjective Programming Models: New Ways in Regional Decision Making,' *Regional Science and Urban Economics*, Vol. 6, No. 3., pp. 253–274, September 1976.

Index of specialized terms

Studies in applied regional science

Vol. 1
On the use of input-output models for regional planning
W. A. Schaffer

This volume is devoted to the use of input-output techniques in regional planning. The study provides a clear
introduction to the essential ideas of input-output analysis.
Particular emphasis is placed on the intricate problems of
data collection at a regional level.

ISBN 90 207 0626 8

Vol. 2
Forecasting transportation impacts upon land use
P. F. Wendt

This reader concentrates on transportation problems in
urban areas. After a survey of model techniques for
analyzing transportation and land use problems, several
new methods in the field of transportation and land-use
planning are presented.

ISBN 90 207 0627 6

Vol. 3
Estimation of stochastic input-output models
S. D. Gerking.

The primary objective of this monograph is to develop a
method for measuring the uncertainty in estimates of the
technical coefficients in an input-output model. This
study also describes three further applications of the
two-stage least squares estimation technique in an input-
output context.

ISBN 90 207 0628 4

Martinus Nijhoff Publishing

Vol. 4
Locational behavior in manufacturing industries
William R. Latham III

Agglomerative economies form a central concept in regional science. Yet an empirical determination of agglomeration advantages has been minimal up to now. To help remedy the situation, this study contains an effort to gauge the order of magnitude of agglomeration advantages, based on a careful inspection of industrial location data.

ISBN 90 207 0638 1

Vol. 5
Regional economic structure and environmental pollution
B.E.M.G. Coupé

This book deals with the ever-increasing problem of pollution. The author has constructed an extensive interregional model for economic activities and pollution.

Coupé's two-region model is used to calculate an equilibrium in terms of production and pollution abatement.

ISBN 90 207 0646 2

Vol. 6
The demand for urban water
P. Darr, S. L. Feldman, C. Kamen

The range of choice for water management can include adjustments to remedy disequilibria through management of the demand side of the market. This volume explores the components affecting demands using combined economic, engineering and social psychological tools and recommends remedies in tariff design to conform to basic economic postulates.

ISBN 90 207 0647 0

Martinus Nijhoff Publishing

Vol. 7
Production systems and hierarchies of centres
J. Gunnarsson

In this study hierarchies of centres are discussed, with
special references to Tinbergen-Bos systems. The author
also uses component analysis to examine whether the
structure of the Swedish system of centres resembles the
structure of different hierarchy models.
Relations between regularities for optimal systems (the size-
distribution of centres and space-functional relations between types of centres) and input-output coefficients as well
as the location of natural resources are studied, and a
quadratic programming model is proposed. This extension
of the problem makes it possible to determine capacities
of plants and systems of centres simultaneously. Thus,
distribution of plant sizes and city sizes may be studied in
the same model.

ISBN 90 207 0688 8

Vol. 8
Multi-criteria analysis and regional decision-making
A. van Delft and P. Nijkamp

The study focuses on the use of multi-criteria methods as
a tool for adequate decision-making. After a discussion
of traditional evaluation techniques (cost-benefit analysis,
for instance) several multi-criteria decision methods are
reviewed. Particular attention is paid to 'concordance
analysis' as it is called. Several variants of this new evaluation method are developed. The operational aspects of
concordance analysis are highlighted particularly in the
case of conflicting views on regional growth and environmental protection.

ISBN 90 207 0689 6

Martinus Nijhoff Publishing

Vol. 9
Economic aspects of regional welfare
C. P. A. Bartels

This study is focused on the analysis of relationships between various economic aspects of regional welfare, especially income distribution and unemployment. A wide variety of methodological questions are explored, with a particular emphasis on their empirical relevance. Methods for describing income distributions concisely and exploring characteristics of regional unemployment series and alternative ways of defining income inequality measures are considered, with a special emphasis on a full explicit integration of normative and positive elements in this definition. Characteristics of unemployment series are also empirically explored. A link is made between income and employment based on an explanatory analysis, and the study concludes with the presentation of a scheme for an integrated regional labour market/income.

ISBN 90 207 0706 X

Vol. 10
Spatial representation and spatial interaction
Ian Masser and P. J. B. Brown

This book is directed towards regional scientists, geographers, urban, regional and transport planners and others with a particular interest in the practical application of methods of spatial analysis. In recent years, the problem of spatial representation has been recognized as being of fundamental importance to the effective application of a wide range of analytical methods. This book draws together for the first time various related pieces of work undertaken in the field of spatial interaction research and sets them in a

Martinus Nijhoff Publishing

general framework within which the problem of spatial representation is viewed as part of the general problem of aggregation. Two kinds of strategy for dealing with this problem in relation to interaction data are outlined in an introductory overview of the field and their practical application is illustrated in subsequent chapters. In the concluding chapter, a number of general themes from the various streams of work are identified and a number of areas defined for further investigation.

ISBN 90 207 0717.5

Vol. 11
Tourism and regional growth
Moheb Ghali, editor

This book is the first systematic empirical study of the growth operations open to a region, the tourist-oriented economy of Hawaii.

The first part is devoted to studying regional growth under resource constraints. This is followed by the study of the region's major exports, their determinants, and projections of their future growth. An econometric model of regional growth is presented in the third part and is utilized to simulate the growth paths of income, employment, migration and unemployment under alternative tourism policies. In the last part of the book, the fiscal implications of the alternative growth paths are derived.

ISBN 90 207 0716 7

Vol. 12
Economies of scale in manufacturing location
Gerald A. Carlino

While agglomeration economies have been of paramount importance in explaining the relative concentration of manufacturing activity in metropolitan or central places, they have not been adequately measured or tested. Most

Martinus Nijhoff Publishing

attempts at measuring agglomeration economies have been
indirect rather than direct, e.g., Marcus' residual growth
proxy or Edel's elasticity of land values with respect to
city size measurement. The present volume suggests a more
direct method of quantifying this rather elusive variable,
through the application of production function techniques.
Also, questions concerning optimal city size and industrial
location are considered.

ISBN 90 207 0721 3

Vol. 13
Urban residential location models
Stephen H. Putman

Research done in the 1970's has yielded a great increase in
our understanding of urban spatial phenomena. A substan-
tial portion of the theoretical and empirical bases of this
work derives from pioneering efforts in urban modeling
done in the 1960's. This book is a collection of excerpts,
commentaries, and evaluation of most of these earlier
modeling projects, focusing on urban residential location,
along with new material which compares and evaluates
them. Anyone, academic, student, or practicing planner,
who is likely to be involved either directly or peripherally
with urban modeling or urban spatial analyses of any sort,
will benefit from a knowledge of the many studies described
in this book.

ISBN 90 207 0785 X